# From Farm to Table

# From Farm to Table

*The Science of Milk and Dairy Products*

ALAN KELLY, PATRICK FOX, AND
TIMOTHY COGAN

OXFORD
UNIVERSITY PRESS

Oxford University Press is a department of the University of Oxford.
It furthers the University's objective of excellence in research, scholarship,
and education by publishing worldwide. Oxford is a registered trade mark of
Oxford University Press in the UK and in certain other countries.

Published in the United States of America by Oxford University Press
198 Madison Avenue, New York, NY 10016, United States of America.



CIP data is on file at the Library of Congress

ISBN 9780197580998

DOI: 10.1093/9780197581025.001.0001

Printed by Integrated Books International, United States of America

MIX
Paper
FSC FSC® C183721

*This book is dedicated to Patrick Fox (PF), coauthor and friend, who passed away before seeing this book published.*

# Contents

# Acknowledgments

The authors would like to thank Jeremy Lewis of Oxford University Press for his openness to our proposal, David Lipp of OUP for support in the production process, and the team at Newgen Knowledge Works for their careful preparation of the manuscript. We would also like to thank David Waldron and Anne Cahalane of University College Cork for their work on the lovely images that bring the products and processes to life, and many friends and colleagues at University College Cork, Teagasc and elsewhere who commented on chapters or provided suggestions. Your inputs are all greatly appreciated.

# Introduction

Milk and dairy products are an important part of the diet and culture of many countries, and have been for millennia. Despite challenges to this today—such as changing dietary patterns, the rise of veganism, and concerns around environmental aspects of milk production—dairy products are a very significant food product category globally and are produced and consumed in almost every country (some statistics on milk consumption are discussed in Chapter 1). They are also produced on a range of scales, from very large companies to small farmhouses and even, in many countries, domestically (e.g., making butter or ice cream at home).

The production of many dairy products from milk has been an art for far longer than it has been a science, and the origin of products like butter and cheese dates back thousands of years, close to the initial domestication of animals like cattle, buffaloes, sheep, and goats. In the intervening centuries, the diversity of dairy products and the sophistication of their production has increased gradually, but arguably the science underpinning their production, characteristics, and safety has really developed only since the start of the 20th century, and still continues to be elucidated.

This book seeks to provide an introduction to the basic science underpinning the way in which milk is transformed into the wide range of dairy products available today. Despite the apparent simplicity of production of many dairy products, milk is a highly complicated material in scientific terms, being a multiphase material (a suspension of fats and proteins in a complex salts and sugar medium) which is biologically active (containing enzymes, bioactive proteins, and a very wide range of nutrients) and inherently unstable, both physically (unhomogenized milk develops a fat-rich cream layer on standing) and microbiologically (raw milk spoils rapidly due to the potential of its nutrients to support the growth of a wide range of microorganisms).

Thus, despite their familiarity (and widespread consumption), milk and dairy products are a very interesting scientific topic, any consideration of which rapidly raises issues of food production, sustainability, chemistry, nutrition, safety, and spoilage, all of which are of considerable interest and subject to wide debate today.

The objective of this book is to provide an entry-level introduction to the world of dairy products, focusing on their chemistry, microbiology, and processing,

*From Farm to Table*. Alan Kelly, Patrick Fox, and Tim Cogan, Oxford University Press.
 DOI: 10.1093/9780197581025.001.0001

at a level intended to be accessible to the non-specialist, while introducing key concepts of the scientific principles involved. Following chapters that introduce the origins and composition of milk and the basic principles of microbiology, particularly as they apply to milk, separate chapters largely relate to specific dairy products or families of products, such as cheese and milk powders.

# 1
# Mammals and Milk Production

Milk is a fluid secreted by the female of all mammalian species—of which there are about 6,500 species in existence today—to meet the complete nutritional requirements of the neonate. This means that milk supplies energy (provided primarily by lipids and the milk sugar, lactose), essential amino acids and fatty acids (essential because mammals cannot produce them themselves), vitamins, inorganic elements (minerals), and water. Milk also serves a number of physiological functions to support the health and development of the neonate, through biological activities of components like immunoglobulins, enzymes, and antibacterial agents.

The nutritional requirements of a neonate differ greatly between species and depend on their maturity at birth, growth rate, and energy requirements, and so the gross composition of milk shows large interspecies differences.

While breast milk is the perfect food for human infants, milk consumption beyond infancy has been part of human civilization for millennia, and the milk of certain domesticated animals, and products made therefrom, are major components of the human diet in many parts of the world. Domesticated goats, sheep, buffalo, and cattle have been used for milk production since about 8000BC.

The milk of only about 50 species has been analyzed in sufficient detail to give reliable compositional data. Milk from the commercially important species—cattle, goat, sheep, buffalo, yak, horse, pig, and human—is, perhaps not surprisingly, the best characterized.

Milk production today is about 800 million tonnes per annum (projected to increase to over 1,000 million tonnes by 2030), about 82% of which is bovine (from cows), 14% buffalo, about 4% caprine (from goats), and about 1% ovine (from sheep), with small amounts produced in certain geographical regions from camels, horses, donkeys, yaks, and reindeer. Milk and dairy products are consumed throughout the world, but are particularly important in Europe, the United States, Canada, Argentina, India, Australia, and New Zealand. The contribution of milk and dairy products to dietary intake varies widely in different regions of the world (see Table 1.1); per-capita, annual consumption of milk in Montenegro (338 kg, the highest level), as estimated by the Food and Agriculture

*From Farm to Table.* Alan Kelly, Patrick Fox, and Tim Cogan, Oxford University Press.
 DOI: 10.1093/9780197581025.003.0001

**Table 1.1** Milk and dairy product consumption per capita in 2020

| Region/country | Per capita consumption (lbs) | Per capita consumption (kgs) |
|---|---|---|
| Montenegro | 744.4 | 338.0 |
| Switzerland | 645.8 | 295.8 |
| Netherlands | 574.8 | 263.2 |
| Denmark | 534.7 | 244.9 |
| Ireland | 526.6 | 241.2 |
| Australia | 511.6 | 234.3 |
| United States | 505.9 | 231.7 |
| France | 437.5 | 200.4 |
| United Kingdom | 436.7 | 198.1 |
| Canada | 355.5 | 161.4 |
| New Zealand | 225.2 | 102.3 |
| India | 147.7 | 67.0 |
| China | 54.6 | 24.8 |

*Source*: Food and Agriculture Organization

Organization in 2020, and the United States (231.7 kg) is several times higher than that in India (67 kg) and China (24.8 kg).

The chemistry and properties of milk have been studied for about 150 years and will be discussed in later chapters of this book. The objective of this opening chapter is to provide a brief summary and overview of the evolution of mammals and lactation.

## The evolution and diversity of mammals

Mammals are thought to have evolved from a type of reptile called synapsids (therapsids), during the Early Jurassic Period, about 200 million years ago. It is estimated that about 80% of all mammalian species have become extinct. It has been suggested that pre-mammalian reptiles secreted a fluid from sebaceous (skin) glands to keep their shell-less eggs moist and prevent microbial infection. The hatchlings would have licked off and ingested some of this fluid, providing them with nutrients. This then became the principal function of such secretions and led ultimately to milk production.

Initially, mammals were small shrew-like creatures, but they have evolved and diversified to occupy niches on land, sea, and air. They range in size from a few grams (pigmy shrews) to 200 tonnes (blue whales) and became dominant after the extinction of the dinosaurs, 60–70 million years ago. Mammals have been successful because the young of most species are born alive and all are supplied with a specially designed food, milk, for the critical period after birth. No other class of animal is so pampered.

Mammals are distinguished from other classes of animals by four criteria:

- They secrete milk to nourish their young.
- They control their body temperature.
- They grow body hair or wool for insulation (even aquatic mammals have some hair).
- They have several different types of teeth which allow them to masticate different types of food.

The word mammal is derived from the Latin word, *mamma*, for breast. The classification and nomenclature of mammals was first proposed by the Swedish naturalist, Carolus Linnaeus, in 1758 and was based initially on differences in morphological characteristics. More scientifically based classification is now possible based on sequences of the genetic materials deoxyribonucleic acid (DNA) and ribonucleic acid (RNA) and on the amino acid sequence of certain proteins, such as caseins.

The Class Mammalia is divided into two sub-classes, Prototheria and Theria. The sub-class Protheria contains only one order, Monotremata, the egg-laying mammals, the duck-billed platypus and four species of echidna, all of which are found only in Australia and New Guinea. Monotremes were the earliest mammals and, presumably, many species have become extinct. They have as many as 200 mammary glands grouped in two areas of the abdomen; the glands do not terminate in a teat and the secreted milk is licked by the young from the surface of the gland. The milk of monotremes is compositionally very different from that of most other mammals.

The young of the sub-class Theria are born alive (such mammals are called viviparous), and they are divided into Metatheria (marsupials) and Eutheria. Marsupials (of which there are about 200 species) do not have a placenta; the young are born alive after a short gestation and are very immature. After birth, they are transferred to a pouch where they mature. Most marsupials are indigenous to Australia, of which the kangaroo, wallaby, and koala are the best known. In marsupials, the mammary glands vary in number and are located within the pouch and terminate in a teat.

About 95% of all extant mammals belong to the sub-class Eutheria (or placental mammals). The developing embryo in such mammals receives nourishment in utero from the placental blood supply and is born at a high, but variable, species-related state of maturity. The number of mammary glands in eutherians varies with the species, from two (human, goat, sheep, horse, and donkey) to four in cattle and water buffalo, 14–16 in the pig, up to 22 in some insectivores. Each gland is anatomically and physiologically separate and is emptied via a teat.

## Classification of the principal modern dairying species

Following the scheme developed by Linnaeus, all forms of life are classified at a number of different levels, which in descending order are domain, kingdom, phylum, class, order, family, genus, and species, so that each family, for example, contains multiple genera, and each genus multiple species. All the principal and most of the minor bovine and buffalo dairying species belong to the Family *Bovidae*, a member of the Order Artiodactyla, which also includes even-toed ungulates, i.e., cloven-hoofed animals such as pigs, goats, and sheep. The *Bovidae* evolved ~18 million years ago; the earliest *Bovidae* fossil (called *Eotragus*) was found in 18-million-year-old deposits in Pakistan. The Order Artiodactyla has three sub-orders: *Ruminantia* (ruminants, to which all major dairying species belong), *Sunia* (pigs and related species), and *Tylopoda* (camels, llama, alpaca, and guanaco). The ruminants are classified into six Families: Tragulidae (chrevrotrains), Moschidae (musk deer), Antilocapridae (pronghorns), Giraffidae (giraffes and okapi), Cervidae (deer), and *Bovidae* (bovines, of which there are 137–138 species in 46–47 genera).

The Bovidae are then divided into six sub-families, of which Bovinae is the most important. The Bovinae are divided into three Tribes, of which the Bovini are the most important from a dairying viewpoint. The Bovini are classified into five genera: *Bubalus* (water buffalo), *Bos* (cattle), *Pseudoryx* (also known as saola, a rare forest-dwelling mammal found in Vietnam), *Syncerus* (African or Cape buffalo), and *Bison* (American and European buffalo). Today, there are about a billion cattle worldwide, of which the two main species are *B. taurus* and *B. indicus*. The latter dominate in Africa and are less efficient producers of milk and meat than *B. taurus*, but are more resistant to heat stress and various diseases and therefore dominate in tropical regions.

Since cattle were first domesticated ~8,000 years ago, they have been bred selectively, especially during the past 200 years, for various characteristics, e.g., health, fertility, docility, and milk or meat production, or both. Today, there are about 800 breeds of cattle, including dairy, beef, or dual purposes breeds. There are ~270 million dairy cows today, of many breeds, the Holstein-Friesian being the principal breed in many countries, representing ~35% of the total. Other important international dairy breeds are Brown Swiss, Jersey, Ayrshire, Guernsey, and Danish Red.

There are about 160 million buffalo worldwide, of which there are two types, river and swamp, found mainly in Southeast Asia, India, and Egypt, with smaller numbers in Bulgaria, Italy, Brazil, and Australia. Depending largely on the region, buffalo are used for milk (and dairy products, such as mozzarella di bufala cheese in Italy), meat or work, or combinations of these.

The global population of goats is about 1 billion, of which about a quarter are dairy goats, mainly in Asia, Africa, and southern Europe. About 80 breeds of goat are recognized, the most popular breed being the Nubian, from Egypt and Ethiopia.

There are also about 1 billion sheep globally, of which about 250 million are used for milk production. Dairy sheep are found in many countries around the world, but the vast majority are in southern Europe, especially France and Spain. About 12 breeds of dairy sheep are recognized, and the East Friesian is considered the best for dairy production.

Camel milk has long been consumed by desert travelers, but intensive dairy farming of camels is increasing in many areas. Dromedary (one-humped) camels yield 4 to 5 times more milk than Bactrian (two-humped) camels; camel milk, which yields good quality fermented milk products, can be used for lactic butter production and is coagulable only by rennet produced from camel stomachs. About 2.85 million tonnes of camel milk are produced annually, principally in Somalia, Kenya, Niger, Mali, Ethiopia, and Saudi Arabia.

The main milk-producing species are shown in Figure 1.1.

## Structure of the mammary gland

The mammary glands of all species have the same basic structure, and all are located external to the body cavity. Milk constituents are synthesized in specialized epithelial cells (secretory cells or mammocytes) from chemicals present in and absorbed from blood. The secretory cells are grouped as a single layer on the surface of spherical or pear-shaped bodies, known as alveoli, which contain a central space, the lumen. Milk is secreted from these cells into the lumen of the alveoli and, when the lumen is full, the cells surrounding each alveolus contract, under the influence of the hormone oxytocin, and the milk is drained *via* a system of ducts into cisterns, which are the main collection points between suckling or milking. These cisterns lead to the outside via the teat canal. A schematic diagram of the mammary gland is shown in Figure 1.2.

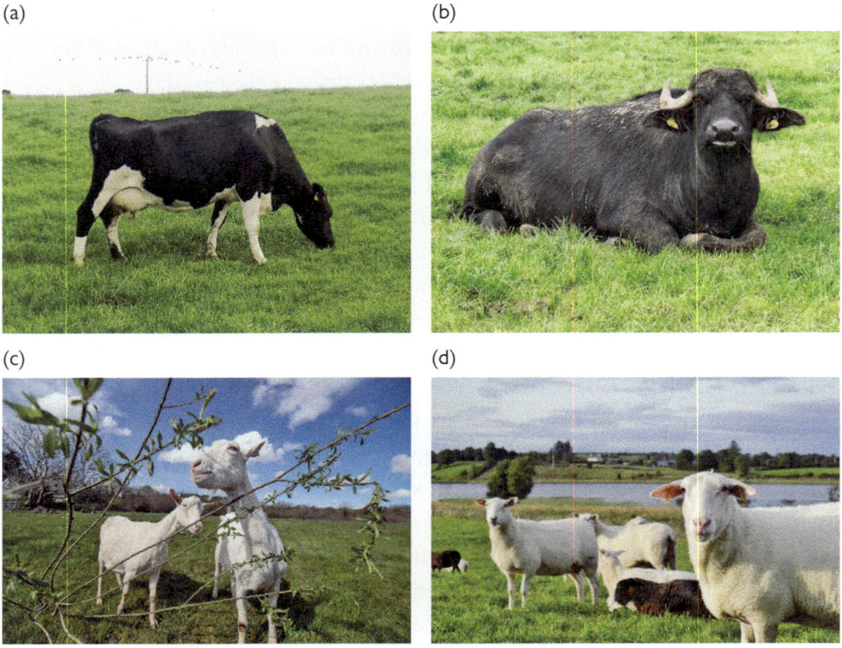


**Figure 1.1** The major milk-producing animals (from top left clockwise): Holstein cow, Buffalo, dairy sheep (by permission of Velvet Cloud yoghurt, Claremorris, Co. Mayo, Ireland) and dairy goats (by permission of St Tola Cheese, Co. Clare, Ireland).

Milk constituents are synthesized from components obtained from the blood; for this reason, the mammary gland has a plentiful blood supply and an elaborate nervous system to regulate excretion.

A different type of cell, called myoepithelial cells, form a "basket" around each alveolus and are capable of contracting on receiving an electrical, hormonally mediated, stimulus, thereby causing ejection of milk from the lumen into the ducts. This stimulus may be triggered by the action of a calf or baby suckling, or in the case of cattle may be a learned reflex initiated by the commencement of milking. The key hormone controlling milk excretion is oxytocin, while the process is inhibited by another hormone, adrenaline, and so milk excretion is difficult in stressed circumstances.

Development of mammary tissue commences before birth, but at birth the gland is still rudimentary. It remains rudimentary until puberty, when very significant growth occurs in some species; in all species, the mammary gland is fully developed at puberty. In most species, the most rapid phase of mammary

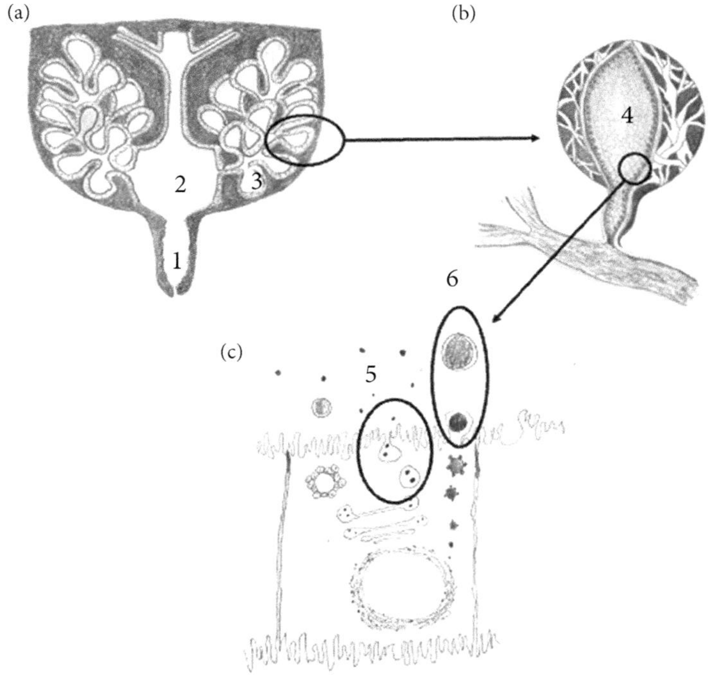


**Figure 1.2** (A) Structure of one quarter of a bovine udder, showing (1) the teat and (2) the udder cistern, in which milk collects after draining through milk ducts from (3) the alveoli. (B) Close-up structure of an alveolus, showing the outer surface lined with (in white) blood vessels and muscle, and the inner surface covered in secretory cells where milk is produced, surrounding (4) the lumen. (C) Secretory cell showing the mechanism of milk secretion, with casein micelles being packaged and released into the milk serum at (5) and milk fat globules being assembled and pushed through the cell membrane, which becomes the milk fat globule membrane, at (6) (image thanks to Anne Cahalane).

gland development occurs during pregnancy and after birth, to enable maximum milk production.

## Structure of the mammary cell

The structure of the secretory cell is essentially similar to that of other eukaryotic (non-bacterial) cells. In their normal state, the cells are roughly cubical, ~10 mm in cross-section. It is estimated that there are ~5 trillion (i.e., $5 \times 10^{12}$)

mammary cells in the udder of the lactating cow. A mammary cell is surrounded by a membrane called the plasmalemma and contains a nucleus where the DNA is found in the form of chromosomes; mitochondria which are responsible for energy generation; ribosomes where proteins are synthesized; and other typical cell organelles within a viscous fluid called the cytoplasm. A simple diagrammatic representation of the cell is shown in Figure 1.2 C.

From studies of the mechanisms of milk synthesis, it has been shown that the specific constituents of milk are synthesized from small molecules absorbed from the blood by protein-based carrier systems that transport small molecules across the cell membrane. The raw materials (substrates) for milk synthesis enter the secretory cell across the basal membrane (facing away from the lumen), and are utilized, converted, and interchanged as they pass inward through the cell. The finished milk constituents are then excreted into the lumen across a cell membrane on the side facing the lumen (the apical membrane). The fat globules are excreted by a process called "blebbing," in which they are pushed through the cell membrane and take part of that membrane with them, forming the milk fat globule membrane (MFGM).

The mammary gland of the mature lactating female of many species is by far the most metabolically active organ of the body. For many small mammals, the energy input required for the milk secreted in a single day may exceed that required to develop the neonate in utero. A cow at peak lactation yielding 45 kg milk/day secretes approximately 2 kg lactose and 1.5 kg each of fat and protein per day, compared to the daily weight gain by a beef animal of 1–1.5 kg/day, 60–70% of which is water. It could be said that a high-yielding dairy cow is subservient to the needs of its mammary gland, to which it must supply not only the precursors for the synthesis of milk constituents but also an adequate level of high-energy-yielding compounds required to drive the necessary biosynthetic reactions, as well as minor constituents (vitamins and minerals).

Some constituents of milk are unique to the mammary gland. These can vary significantly between species (e.g., most milk proteins and lipids), while some are uniquely found in the mammary gland but do not differ between species (e.g., lactose). Some are found in other parts of the mammalian body (e.g., some whey proteins in blood), and some are generic (e.g., vitamins and minerals).

## The utilization of milk

Since young mammals are born at a different stage of maturity and with different nutritional requirements, the milk of each species is designed to meet the requirements of the neonate of that species. Milk is intended to be consumed unchanged by the young suckling its mother. However, humans have consumed the milk of other species for at least 8,000 years.

Milk is often described as the most "nearly perfect" food; although this is true only for the young of the producing or closely related species, the milk of all species is a nutrient-rich and well-balanced food. Milk is also very susceptible to the growth of microorganisms that cause spoilage. To counteract this, a range of products which are more stable than milk have been developed; some of these date from ~4000BC and have evolved desirable characteristics, in addition to their nutritional value. Today, several thousand food products are produced from milk, which fall into the following principal groups: liquid/beverage milk, cheese, milk powders, concentrated milks, fermented milk products, butter (some of which is produced from cream/fat obtained as a by-product in the manufacture of other products), ice cream, infant formula, creams, protein-rich products, and lactose. Some of these groups are very diverse (for example, at least 1,000 varieties of cheese have been described) and will be described in later chapters of this book.

## Factors affecting the composition of milk

The concentration of the principal constituents varies widely among species: the levels of lipids range from 2%–55%, that of proteins, 1%–20%, and lactose 0%–10%, mainly reflecting the energy requirements (in the case of lipids and lactose) and growth rate (mainly proteins) of the neonate. The concentrations of the minor constituents also vary widely. The composition of the milks of a number of mammalian species is compared in Table 1.2.

Within any species, the composition of milk varies among individual animals, between breeds, with the stage of lactation, feed, health of the animal, and many other factors. The fat content of bovine milk shows large inter-breed differences, and within any breed, there is also a wide range of protein content in milk from individual animals.

The composition of milk, and even the profile of its constituents, changes markedly during lactation, especially during the first few days post-partum. The first milk secreted after birth, called colostrum, is very high in biologically active and protective proteins, such as immunoglobulins.

The composition of milk remains relatively constant during mid-lactation but changes considerably again in late lactation, reflecting a process called involution of the mammary gland tissue, when milk production becomes curtailed (for dairy cows, it often corresponds to advanced pregnancy), and the greater influx of blood constituents, such as certain whey proteins. This can sometimes result

**Table 1.2** Composition (%) of milks of selected species

| Species | Fat | Protein | Lactose | Ash | Total solids |
|---|---|---|---|---|---|
| Cow | 3.7 | 3.4 | 4.8 | 0.7 | 12.7 |
| Human | 3.8 | 1.0 | 7.0 | 0.2 | 12.2 |
| Sheep | 7.4 | 4.5 | 4.8 | 1.0 | 19.3 |
| Goat | 4.8 | 2.9 | 4.1 | 0.8 | 12.3 |
| Buffalo | 7.4 | 3.8 | 4.8 | 0.8 | 16.8 |
| Pig | 6.8 | 4.8 | 5.5 | 1.0 | 18.8 |
| Camel | 5.4 | 3.8 | 5.2 | 0.7 | 15.0 |
| Horse | 1.9 | 2.5 | 6.2 | 1.2 | 11.2 |
| Elephant | 33.1 | 10.9 | 0.3 | 1.4 | 47.6 |
| Whale | 40.9 | 11.9 | 1.3 | 1.4 | 55.0 |

in changes in the processing characteristics of milk, such as impaired cheese-making characteristics.

In some countries, dairy cows calve throughout the year, and so milk collected for dairy processing is a mixture of that from cows at different stages of lactation, such that lactational changes in milk properties are not a major consideration for dairy processors. However, in some countries, such as Ireland and New Zealand, calving is highly seasonal and is concentrated into a short time period corresponding to spring (February–March and August–October in the northern and southern hemispheres, respectively). The reason for this is to take advantage of the availability of grass as a plentiful and healthy feed for cattle in outdoor grazing. One consequence of such seasonality is that the processing characteristics of milk can vary at certain times of year, for example when synchronized calving leads to a high proportion of late lactation milk in the supply. The dairy industry in such countries undergoes pronounced peaks and troughs in milk supply, with summer supply sometimes being 10 times that of winter, which has implications both for the operation of milk processing facilities (e.g., factories operating at very reduced capacity in winter) and the types of products produced (e.g., a preference for long-life products such as powders and hard cheese that can be produced in large quantities in summer).

Because some of the raw materials for synthesis of milk components are obtained from the cows' diet, which has a significant effect on the levels and types of milk constituents. Notably, a diet dominated by grass leads to high levels of carotenoids in milk, which gives a yellow-orange color to high-fat dairy products such as butter and cheese. The levels of saturated and unsaturated fats in the diet also influence the levels of such fats in dairy products, which has significant implications for properties such as the hardness of butter. In addition, inclusion of specific feedstuffs like those derived from fish oils can increase the levels of specific types of fat that are perceived as healthy, such as omega-3 and omega-6 fatty acids.

Finally, the health of the cow has a major influence on the production of milk, with mastitis (caused by bacterial infection of the udder) resulting in profound changes in the levels of proteins, lactose, enzymes, and minerals in milk, all of which can negatively impact on the production of dairy products. For this reason, a major parameter used to evaluate the quality of milk for processing is the level of white blood cells in milk (the so-called somatic cell count), as these become concentrated in the udder to fight the infection. Milk with a high somatic cell count indicates the presence of cows suffering from mastitis and is typically not allowed for processing; the infected quarter is generally treated with an antibiotic and the milk collected separately until all the antibiotic has been excreted, which can, depending on the antibiotic, take six to eight milkings.

Overall, milk is a complex biological fluid, and many of its properties are variable, due to a number of considerations. Before discussing the ways in which it is converted into dairy products, and to understand the changes that take place during such processing, it is important to explore in detail the properties of the principal milk constituents, which will be the focus of the next chapter.

# 2
# The Chemistry of Milk Components

Milk is a very complex system. Its principal constituents are, in decreasing order of concentration, water, carbohydrates (mainly lactose), lipids, and proteins, with low levels of organic and inorganic salts and vitamins. The composition of milk shows considerable variability between species, as do the characteristics and properties of the principal constituents. This chapter presents brief overviews of the properties of these constituents, as they are important in understanding the manufacture and properties of all dairy products.

## Water

Water is the principal constituent in the milk of most species (bovine and human milk both contain around 87% water by weight). In addition to meeting the requirement of the neonate for water, the water in milk serves as a solvent for milk salts, lactose, and proteins and affects their properties and stability. It also controls the rate of many chemical reactions, the breakdown of proteins and lipids, activities of enzymes, and microbial growth, and thus greatly affects the stability of milk and milk products. Milk can be very effectively preserved by reducing the level or availability of water by dehydration (as in powder, cheese, and butter production) or by adding sucrose (e.g. in the production of sweetened condensed milk).

## Carbohydrates

### Lactose

The principal carbohydrate in the milk of most species is lactose, which is composed of one molecule of glucose bonded to one molecule of galactose; such a sugar is called a disaccharide. The concentration of lactose in milk is characteristic of each species and varies from 0 to ~10%: around 4%–5% in the milk of

*From Farm to Table*. Alan Kelly, Patrick Fox, and Tim Cogan, Oxford University Press.
 DOI: 10.1093/9780197581025.003.0002

cows, goats, and sheep; around 7.0% in that of humans; and less than 1.0% in the milk of polar bears and seals. Milk is the only known source of lactose. It was first isolated in about 1780 by Carl Scheele, and its chemistry and important physico-chemical properties have since been described thoroughly.

Lactose serves as a ready source of energy for the neonate (it provides ~30% of the calories in bovine milk), and its principal function is as an energy source. However, since lipids are 2.5 times as energy dense as lactose, when a highly calorific milk is required, e.g., by animals that live in a cold environment, the fat content of the milk increases more than that of lactose.

Lactose is synthesized in the mammary cells from two molecules of glucose absorbed from blood, one of which is converted to galactose, which is then joined by a chemical bond to another molecule of glucose. The mechanism by which this happens, interestingly, requires the presence of the milk protein α-lactalbumin; there is a positive correlation between the concentrations of lactose and this protein in milk. The synthesis of lactose in the mammary gland causes water to be drawn into the region of the cell where lipids and proteins are synthesized (the Golgi vesicles), thereby diluting their concentration, and so it also plays a key role in controlling the yield and volume of milk produced.

The properties of lactose are generally similar to those of other sugars, but it differs in some technologically important respects.

First, lactose is a reducing sugar, which means in chemical terms that it can donate electrons to other molecules. In practical terms, the main significance of this is that it can undergo reactions with the amino group of amino acids (called the Maillard reaction, after the French chemist Louis-Camille Maillard, who first studied them) at high temperatures. The Maillard reaction leads to the development of brown colors in heated milk or products (such as bakery products) to which milk is added, and can also result in the production of off-flavors. It is referred to as (non-enzymatic) browning, and contributes positively to the flavor and color of many foods, e.g., the crust of bread, toast, and deep-fried products. The effects in dairy products can of course be negative if a brown color is undesirable.

Second, lactose exists as two isomers (forms of a molecule which have the same chemical composition but in which the atoms are arranged differently), called alpha (α) and beta (β). These are interchangeable forms, and the process by which one changes to the other is called mutarotation. The α and β isomers of lactose have very different properties, of which the most important are solubility in water (70 g/L and 500 g/L, for α and β, respectively, at 20°C). The solubility of α lactose is more temperature-dependent than that of β lactose, and it is the more soluble form above 94°C.

When lactose becomes insoluble or is present in highly concentrated systems like powders, it forms crystals, the properties of which have major significance on many dairy products. The isomers differ in their ability to form such crystals, due to the differences in their solubility; α lactose crystallizes at temperatures below 94°C and so is the usual form of lactose found in commercial products, while β lactose may is found if crystallization occurs at temperatures above 94°C.

During drying of lactose-rich materials, any lactose that has not been crystallized forms an amorphous unstructured material, called a glass, which is stable if the moisture content of the powder is low. However, if the moisture content exceeds ~6%, the lactose crystallizes and, if a certain type of lactose crystal forms, the products can readily absorb water and solidify into hard-to-handle interlocking crystalline masses or lumps, a process called caking. Lactose, especially the α form, forms large tomahawk-shaped crystals, which may cause problems (such as caking and gritty texture) in lactose–rich dairy products, e.g., skim milk and whey powders, unless crystallization is controlled.

These problems can be avoided by adequate crystallization of lactose before drying, which involves a holding step during the drying, or by using packaging that prevents the uptake of water. If the crystallization process is carefully controlled in this way, the lactose crystals will not absorb water, which renders it very suitable for applications like icing sugar blends.

Lactose has a low level of sweetness; it is about 16% as sweet as sugar (sucrose) and hence has limited value as a sweetening agent. It is, however, a useful bulking agent when excessive sweetness is undesirable; for this reason, and its ability to form solid structures, because of its crystallization behavior, it is used in some pharmaceutical tablets.

In the manufacture of fermented dairy products, such as cheese and yoghurt, lactose serves as a carbon and energy source for lactic acid bacteria, which produce lactic acid, an essential compound in the production of these products.

## Lactose intolerance

Mammals are unable to absorb lactose from the small intestine, so they secrete an enzyme called β-galactosidase (commonly known as lactase) to break it down to glucose and galactose, which can be absorbed into the blood. The young of most mammals secrete adequate β-galactosidase but, as they age, they produce less of this enzyme and eventually are not able to hydrolyze significant dietary intakes of lactose. Lactose then enters the large intestine, where it is metabolized by bacteria with the production of gas, which causes diarrhea, cramps, and flatulence. Humans typically enter the lactose-intolerant state at 8–10 years of age, but in some populations a mutation that occurred thousands of years ago has enabled continued production of the enzyme, and hence the ability to tolerate dairy products persists into adulthood.

The frequency and intensity of lactose intolerance/malabsorption today varies widely among populations from ~100% in Southeast Asia to ~5% in Northwestern Europe. Most northern Europeans and some African tribes continue to secrete adequate β-galactosidase throughout life, because, in evolutionary terms, the consumption of milk has selected for this characteristic over thousands of years.

Interestingly, some of the oldest dairy products, such as fermented milk and cheese, have reduced levels of lactose, due to conversion to lactic acid and/or removal in whey, and so their development may have helped individuals with lactose intolerance consume and avail of the other nutritional benefits of these dairy products.

A study published in 2022 examined the relationship between historical variations within European populations in their ability to digest lactose (termed lactase persistence, LP) and evidence of milk usage through fat residues on over 7,000 pottery samples from more than 550 archaeological sites. Notably, there did not seem to be a strong association between ancient dairy consumption and modern lactase persistence, suggesting that factors such as famine and exposure to pathogens may have played roles in driving the increasing incidence of LP in Europe.[1]

Lactose-intolerant people usually avoid consuming milk, but the problems caused by lactose intolerance may be avoided by breaking the lactose down into glucose and galactose using purified lactase, either in the plant during processing or by the consumer using lactase preparations purchased from pharmacies.

## Oligosaccharides

Simple sugars include monosaccharides like glucose, or sugars like lactose that comprise 2 monosaccharide molecules bonded together; sugars comprising 3 or 4 monosaccharides are called, tri- or tera-saccharides, respectively, while polysaccharides such as starch comprise hundreds or thousands of monosaccharide building blocks assembled into much more complex structures. Complex sugars that contain 5–10 monosaccharide units are called oligosaccharides (OSs). Almost all the OSs in milk have lactose at one end of the molecule (chemically, the reducing end), linear or branched molecular structures, and frequently, uncommon sugar molecules like fucose and N-acetylneuraminic acid. Fucose occurs widely in tissues of mammals and other animals where it serves many functions, but its significance in the oligosaccharides in milk is not clear.

The milk of all species contains OSs, and the highest levels occur in the milk of monotrenes (egg-laying mammals), marsupials, marine mammals, humans, elephants, and bears. With the exception of humans, the milk of these species contains little or no lactose, and OSs are the principal carbohydrates. Human milk contains ~130 different OSs, at a total concentration of ~15 g/L; they are considered to be important for neonatal brain development. The concentration of OSs is higher in colostrum (the first milk produced after birth) than in milk.

There is currently significant interest in producing OSs similar to those in human milk for addition to infant formula, due to their apparent importance. Possible methods for doing this include fermentation, chemical synthesis, or recovering and concentrating OSs from bovine milk, whey, or permeate from the filtration of milk or whey to recover proteins.

The significance of oligosaccharides is not clear, but some aspects may be significant. For example, they are not hydrolyzed by β-galactosidase, and enzymes required to hydrolyze the sugars they contain are not secreted in the intestine. So, the oligosaccharides function as undigested soluble fiber and prebiotics, which affects the microflora of the large intestine (i.e., the complex populations of bacteria present). It is claimed that they prevent the adhesion of pathogenic bacteria to the intestine by acting as "decoys," as they have a similar structure to cell structures the bacteria would otherwise adhere to. Galactose and, especially, N-acetylneuraminic acid, are important for the synthesis of glycolipids and glycoproteins, important compounds in brain and nerve development, which may be a key function of the OSs.

Sugar molecules in milk may also be found as part of the structure of proteins and lipids, in which case the resulting molecules are referred to as glycoproteins or glycolipids, respectively. Some milk proteins, especially κ-casein, are glycoproteins, while glycolipids are found in the milk fat globule membrane (MFGM).

## Lipids

Lipids (commonly called oils or fats, depending on whether they are liquid or solid, respectively, at ambient temperature) are those constituents of tissues, biological fluids, or foods that are insoluble in water but are soluble in organic solvents such as diethyl ether or chloroform. Historically, the fat of milk was regarded as its most valuable constituent, and the value of milk was based largely, or totally, on its fat content since butter was the main dairy product produced.

The level of fat in milk varies widely between species, ranging from ~2% to over 50%. Fat levels are lowest in the milk of horses and donkeys (<2%), and vary from 3% to 5% in human, bovine, and goat milk, with over 7% being found in sheep milk.

The vast majority of lipids are formed by a reaction between a simple molecule called glycerol and carboxylic or fatty acids (FAs). Fatty acids typically contain chains of carbon atoms, each bonded to 2 neighboring carbon atoms plus 2 hydrogen atoms (for this reason, such molecules are often called hydrocarbons). At one end of the molecule, the carbon is bonded to a third hydrogen atom, while at the other end there is an acidic group containing another carbon atom plus oxygen atoms.

Glycerol is a much simpler molecule, containing 3 chemical groups called hydroxyls, which contain an oxygen and a hydrogen atom (an OH group, characteristic of all alcohols). The acidic end of a fatty acid can bind to one of these hydroxyl groups, by the formation of a strong chemical bond called an ester bond. Molecules formed from the reaction of an acid with an alcohol (glycerol), including lipids, are called esters.

If 1 fatty acid molecule is bound to 1 glycerol molecule, the complex is called a monoglyceride while, if 2 or 3 fatty acids bind to a glycerol molecule, the result is a di- or tri-glyceride, respectively.

Lipids are commonly divided into 3 classes:

***Neutral lipids***: these are typically mono-, di-, and tri-glycerides and are the dominant class of lipids in all foods and tissues, representing 98.5% of total milk lipids. Most FAs in milk are not in a free form but are found in triglycerides. Each glycerol molecule has three points at which fatty acids can be attached, referred to as Sn1, Sn2, and Sn3, with Sn2 being the middle one. As well as the constituent FAs, the position of the FAs in triglycerides affects their melting point and their textural properties.

***Polar (charged) lipids***: many of these contain phosphoric acid, a nitrogen-containing compound (choline, ethanolamine, or serine) or a sugar/oligosaccharide, all of which give the molecule a charge and make it polar, and able to interact readily with water. The hydrocarbon chains of fatty acids, on the other hand, are non-polar and do not attract water. Mono- and diglycerides are also polar or amphiphilic (having regions of the molecule, the fatty acids, that are hydrophobic and repel water and regions that are hydrophilic and attract water, like the charged hydroxyl regions). Although present at low levels (~1% of total milk lipids), these amphiphilic polar lipids are critical for the ability of milk to contain the rest of the lipids as an emulsion of droplets (called milk fat globules) and are concentrated in the MFGM which surrounds and protects these globules and ensures their physical and chemical stability.

***Miscellaneous lipids***: these are a heterogeneous group of compounds which are unrelated chemically to neutral or polar lipids or to each other. This group includes cholesterol, carotenoids, and the fat-soluble vitamins, A, D, E, and K. Carotenoids are pigments (yellow, orange, red) that are responsible for the color of butter and cheese; some consumers prefer highly colored cheese, which is produced by adding a carotenoid-containing extract from annatto beans. Some carotenoids are converted to vitamin A in the liver.

## Types of fatty acid in milk

As mentioned above, FAs are a type of molecule called carboxylic acids, the core of which is a chain containing 4–26 carbon atoms. They are typically referred to as short-chain when they contain 4–8 carbon atoms, medium-chain (10–16 carbons) or long-chain (more than 16 carbons). The chemical bonds between the carbon atoms in a fatty acid can be either a single bonds or a double bond (where the term relates to the number of electrons shared in the bond) and FAs may be either saturated (only single bonds between the carbon atoms) or unsaturated (containing 1–6 carbon-carbon double bonds, called mono- or polyunsaturated, if there are one or more double bonds, respectively). Animal-derived lipids tend to contain more saturated fats, while vegetable-derived oils contain more unsaturated and polyunsaturated FAs. There are two types of unsaturated FAs depending on the arrangement of hydrogen atoms around the double bond, called cis and trans.

Most FAs contain an even number of carbon atoms because they are synthesized from, and elongated by adding, a 2-carbon compound called acetyl-CoA, on each cycle of their synthesis. The shorthand nomenclature for a fatty acid indicates the number of carbon atoms followed by a colon and the number of double bonds, e.g., $C_{8:0}$ (octanoic acid, also known as caprylic acid, a saturated fatty acid, which contains 8 carbon atoms and no double bonds).

Some fatty acids contain hydroxyl (OH) groups along their chain. Called hydroxy fatty acids, these are present at low levels in milk fat, but are important because on heating they are converted to compounds called lactones, which give desirable flavors to milk fat, considered the premium cooking fat. Other rare fatty acids include keto acids, which are important flavor precursors in milk, as they are converted to highly-flavored compounds that contribute to the flavor of cheese.

The melting point (MP) of FAs—the temperature at which they liquify from a solid crystalline state—is a key property for the texture of high-fat foods like butter and cheese. Their MP increases and their solubility in aqueous solvents decreases with size (molecular mass) and with the number of double bonds in the FAs. Fats rich in unsaturated FAs, such as vegetable oils, are liquid at room temperature, while those with more saturated fats, such as animal fats, are solid. Butyric acid (also known as butanoic acid), a short-chain FA found in milk, has 4 carbon atoms and melts at around −8°C: above this temperature, it is liquid, but below that point it forms crystals. For comparison, capric acid with 12 carbons has a melting point of 31°C, while palmitic acid (the major fatty acid in palm oil) has 16 carbons and a melting point around 62°C. Oleic acid (the major FA found in olive oil) has a melting point of 14°C because, although it has 18 carbon atoms, it has one double bond.

Milk lipids are chemically similar to lipids from other sources, but milk contains a very wide range of FAs—up to 400 FAs have been reported in bovine milk lipids, most of them in trace amounts. The milk lipids of ruminants are unique in that they are the only natural lipids that contain butyric acid. They also contain substantial amounts of the saturated fatty acids: hexanoic (6 carbons, also known as caproic acid), octanoic (8 carbons, caprylic acid) and decanoic (10 carbons, capric acid). The only other sources of these are coconut and palm kernel oil. The short- and medium-chain fatty acids are water soluble and volatile, with a strong aroma and taste.

The fatty acids in milk fat are obtained from three sources:

- bacterial biosynthesis in the rumen;
- synthesis in the mammary gland;
- diet.

Ruminant milk fats contain low levels of polyunsaturated fatty acids (PUFAs) because those consumed in the diet are converted by bacteria in the rumen to saturated forms. This can be prevented by encapsulating (coating) dietary PUFAs or PUFA-rich sources in protein that has been cross-linked (by formaldehyde) or cross-linked crushed oilseeds. Butter produced from PUFA-enriched milk has improved spreadability and may also have improved nutritional qualities.

Incomplete hydrogenation by the rumen bacterium *Butyrivibrio fibrisolvens* results in the formation of conjugated linoleic acid (CLA)—also called rumenic acid—which has been reported to have a number of health benefits, including anticarcinogenic properties. The formation of CLA in milk and its nutritional benefits have been the subject of considerable research in recent years.

## Degradation of lipids

Food lipids are susceptible to two forms of deterioration that can have negative implications for food quality and taste: chemical oxidation of the double bonds leading to oxidative rancidity (giving sharp and unpleasant tastes and aromas) and hydrolysis by enzymes called lipases (lipolysis), which leads to hydrolytic rancidity. Oxidation involves a very complex set of chemical reactions catalyzed by metal ions, especially copper.

Milk contains a lipoprotein lipase (LPL), which is normally inactive because it is separated from the triglyceride substrates on which it would otherwise act by the MFGM. However, when this membrane is damaged, lipolysis and rancidity (due to lipolysis) ensues rapidly. When milk lipids are hydrolyzed by milk LPL, short- and medium-chain FAs are released. These are major contributors to flavor, which may be desirable or undesirable, depending on the product. Hydrolytic rancidity can be a very serious problem in raw milk and in some dairy products.

However, a low level of lipolysis is desirable in all types of cheese, especially in blue cheeses, in which the principal lipases are those secreted by the mold, *Penicillium roqueforti*. The free FAs are converted to methyl ketones, the principal flavor compounds in such cheese. The characteristic piquant flavor of some other cheeses, e.g., Pecorino Romano, is due to the release of short- and medium-chain FAs, mainly by an enzyme called pregastric esterase added to the milk at the beginning of cheese manufacture. Other important derivatives of FAs are called secondary alcohols, lactones, esters, and thioesters, and these are also important flavor compounds in cheese.

## The milk lipid emulsion

Lipids are insoluble in water or aqueous systems. When mixed, lipid and water (or aqueous solvent) form distinct layers, with a destabilizing force, interfacial tension, between and separating the layers. Lipids can be dispersed in an aqueous system by vigorous agitation, but when agitation ceases, the droplets of lipid coalesce quickly into a single mass (i.e. they separate into oil and aqueous layers or phases), driven by the need to minimize the interfacial area between oil and water.

This interfacial tension can be reduced by using surface-active agents (such as emulsifiers or detergents), which bind to dispersed droplets containing the lipids. Natural emulsifiers include proteins, phospholipids, and mono- and di-glycerides, which are all amphiphilic molecules that interact at an interface between hydrophobic (oil) and hydrophilic (water) components, because the structure of such molecules contains both hydrophobic and hydrophilic regions. Emulsifiers are widely used as food ingredients. When the interfacial tension is reduced in this way, the droplets of lipid will remain discrete, although they will rise to the surface (as cream), owing to the lower density of lipids compared to that of water or milk serum.

In milk, the lipids are dispersed in the milk serum as globules with a diameter in the range <1 μm to ~20 μm (1 μm = 0.001 mm), with a mean size of 3–4 μm for bovine milk, to form an oil-in-water (o/w) emulsion, so called because the oil is dispersed in droplets in the water phase. A schematic diagram of the structure of a bovine milk fat globule is shown in Figure 2.1.

In the mammary cells, the fatty acids (from the sources described above) and monoglycerides (from blood lipids) are synthesized into triglycerides, which then form into globules and are released into the cell fluid or cytoplasm. The globules are stabilized by the MFGM, a complex layer of proteins and phospholipids derived from the material that comprises the wall of the cell. Since the stability of the milk emulsion is critical in most dairy products, the structure and stability of the MFGM has been the subject of research for more than 100 years.

The 8 main proteins in the MFGM have been isolated and characterized. However, more sensitive analytical techniques indicate that there are also 120 minor proteins, which serve various biological functions in the mammary cell. Many of the enzymes in milk are also concentrated in the MFGM. Some of the MFGM is shed during storage of milk, especially when agitated, and forms small structures called vesicles (or microsomes) in the skimmed milk.

The MFGM may be damaged by agitation, homogenization, whipping, or freezing, and this may lead to hydrolytic rancidity and the release of non-globular

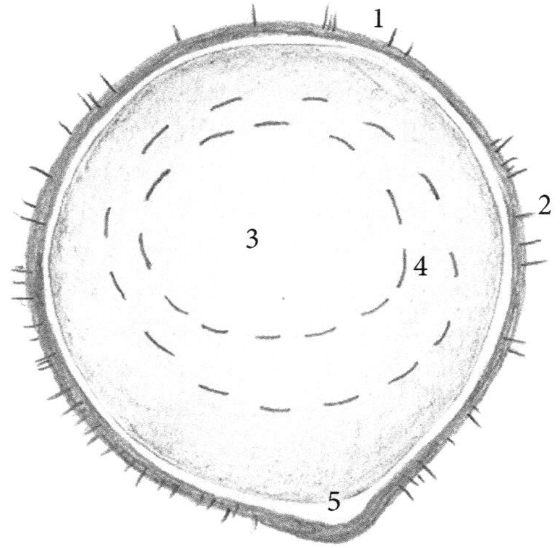


**Figure 2.1** Schematic figure of a milk fat globule in refrigerated milk. The globule is surrounded by a milk fat globule membrane (MFGM, 1), within which are embedded proteins and enzymes (2). The interior of the globule (3) is filled with milk fat, mainly triglycerides, some of which form crystals in cooled milk, which often form concentric shells (4). The MFGM sometimes entraps pools of cell fluid in what are called cytoplasmic crescents (5) which can also contain other components like enzymes (image thanks to Anne Cahalane).

(free) fat, which may cause cream plugs, oiling-off in coffee and tea, and poor wettability of milk powder. In butter making, the MFGM is stripped from the fat globules by extensive agitation (usually of cream), a process referred to as churning; this free fat coalesces and is kneaded (worked) to form butter, a water-in-oil emulsion (see Chapter 8). During churning, the MFGM partitions into the aqueous phase, the buttermilk.

On standing, the fat globules in bovine milk rise to form a cream layer, due to the difference in specific gravity between the fat and aqueous phases, but this cream layer is dispersed readily by gentle agitation. The rate of creaming is influenced by a number of factors, such as the size of the fat globules, the density difference between them and the surrounding medium, and gravity (or centrifugal force applied), all of which increase the rate of separation of globules, and the viscosity of the surrounding medium, which is inversely proportional to the rate of separation, as globules rise more slowly through more viscous fluids.

This can be expressed in an equation called Stokes' law:

$$v = 2r^2\,(\rho 1 - \rho 2)\,g\,/9\eta$$

where v is the rate at which the fat globules rise, r is the radius of the fat globules, $\rho 1$ and $\rho 2$ are the specific gravity of the continuous (skim milk) and dispersed (fat globules) phases, respectively, g is the acceleration due to gravity, and $\eta$ is the viscosity of the continuous phase.

Based on the typical values for these parameters in milk, one would expect a cream layer to form in bovine milk in about 60 hours, but, in fact, a cream layer forms in about 30 minutes. The faster than expected rate of creaming is due to the aggregation of fat globules, aided by a specific protein (an immunoglobulin called cryoglobulin, which precipitates onto the fat globules when the milk is cooled). The clusters of globules behave as a single entity, which has a larger radius and hence greater speed of separation. Creaming can be prevented by homogenizing the milk, which reduces the size of the globules and prevents the action of the cryoglobulins. The fat globules in the milk of some other species, e.g., buffalo, sheep, goat, horse, and camel, do not agglutinate, because these milks lack cryoglobulins, and so these cream more slowly than bovine milk. Human milk does cream, but does not display the faster separation at refrigeration temperatures seen for bovine milk, due to differences in the immunoglobulins present.

Traditionally, the fat was removed from milk by natural (gravity) creaming, and this is still used to adjust the fat content of milk for making some cheese varieties, e.g., Parmigiano Reggiano, but the removal of fat from milk is now usually accomplished by centrifugal separation, which is very efficient, essentially instantaneous, and continuous (as will be discussed in Chapter 5).

## Milk Proteins

The properties of milk and of most dairy products are affected more by the proteins they contain than by any other constituent. The milk proteins have many unique properties and, because of this and their technological importance, they have been studied extensively and are probably the best characterized food protein systems.

Research on milk proteins dates from the early 19th century and by the 1880s, three families of milk proteins—caseins, lactalbumins and lactoglobulins—had been identified. Casein was defined as the milk protein that precipitated at a pH value of 4.6 (a measure of acidity). The pH value reflects the level of hydrogen ions, the main agent of acidity, in a solution and ranges from 0–14, with 0 being the most acidic and 14 being the most basic or alkaline. A value of pH 7 is neutral and means there are the same number of $H^+$ as hydroxyl ($OH^-$) ions. The pH of milk is typically around 6.7, close to neutral, and the addition of acid, or production of lactic acid from fermentation, reduces the pH; gelation or precipitation occurs at pH 4.6. Lactalbumins and lactoglobulins, which remain soluble at pH 4.6 and are the major proteins in whey, are now called whey proteins.

The caseins include a family of four proteins called $\alpha_{s1}$-, $\alpha_{s2}$, β- and κ-casein. A fifth type, the γ-caseins, comprises a group of three protein fragments produced by the breakdown of β-casein by an enzyme called plasmin that is naturally found in milk. This can happen in the mammary gland, and so in raw milk some of the β-casein has already been converted to γ-caseins. The caseins all have related structural features and somewhat unusual properties, as will be discussed later.

In addition to the caseins and whey proteins, milk contains two other nitrogenous materials, proteose peptones (mainly breakdown products of caseins produced by the enzyme plasmin) and non-protein nitrogen (NPN), which were first recognized in 1938. Thus, the protein fraction of milk consists of caseins, lactalbumins, lactoglobulins, proteose peptones, and NPN, which represent approximately 78%, 12%, 5%, 2%, and 3%, respectively, of the nitrogen-containing compounds in bovine milk.

### The structure of proteins

Proteins are polymers of amino acids, and different proteins can have very different levels of complexity in their structure, depending on the amino acids they contain. The simplest level of structure, called the primary structure, is their amino acid sequence. Twenty different amino acids are found in proteins, and each protein can contain from 100 to over 1,000 of them (the caseins contain 150–220 amino acids).

The exact order in which amino acids exist in a protein, and the interactions between them (whether their chemical properties cause them to attract or repel neighboring amino acids, for example) cause specific stretches of the amino acid chain to adopt distinct shapes, including spiral structures such as helices (a very common structural element of many proteins is called the $\alpha$-helix); this type of local structure is called the secondary structure of a protein.

When interactions occur between more distant parts of a protein molecule, this can cause more significant folding of the protein into complex three-dimensional shapes, which is called the tertiary structure of the protein. Disulfide bonds, for example, between pairs of the amino acid cysteine within the protein cause it to fold so as to allow these bonds to form; this is one of the key reactions in the formation of the 3-D structures of proteins.

The final level of protein structure is called quaternary structure and involves different protein molecules combining to form a single larger molecule, e.g., hemoglobin, the oxygen-carrying molecule in blood, consists of a complex of 4 protein molecules.

The key function of milk proteins is of course as a source of the essential amino acids in the diet: valine, isoleucine, methionine, phenylalanine, lysine, leucine, tryptophan, and tyrosine. The milk proteins are hydrolyzed to amino acids in the gastrointestinal system by a series of enzymes called proteinases (pepsin, trypsin, chymotrypsin) and peptidases (which break down the peptides produced by the proteinases). They are then absorbed into the blood and transported to the liver, where they are catabolized or used in the synthesis of other proteins.

## Properties of casein and whey proteins

The properties of caseins and whey proteins differ in a number of significant ways, as summarized in Table 2.1. Firstly, the caseins are, by definition, insoluble at pH 4.6, whereas the whey proteins are soluble under these conditions. So-called isoelectric precipitation of casein (by adjustment of the pH to 4.6) is exploited in the production of caseins and caseinates, as well as some fermented milk products, e.g., cottage cheese and acid-coagulated cheeses, as will be discussed in later chapters.

Secondly, caseins are also coagulable following specific, limited proteolysis by enzymes called rennets, whereas the whey proteins are not. This property of the caseins is exploited in the production of rennet-coagulated cheese (~75% of all cheese) and rennet casein.

**Table 2.1** Key properties of casein and whey proteins

| | Caseins | Whey proteins |
|---|---|---|
| Form in milk | Large aggregates (micelles) | Single molecules or pairs thereof |
| Impact on milk properties | Mainly responsible for white color | Little in raw milk |
| Impact of high temperature | Very little | Denature and interact |
| Impact of rennet | Form a gel | None |
| Impact of acidification | Insoluble at pH 4.6 | Soluble at pH 4.6 |
| Key products | Form the basis of cheese and yoghurt; can be isolated to give a range of products | Whey protein ingredients for nutritional or structure-forming applications |

Thirdly, caseins are much more heat stable than the whey proteins. Milk at its normal pH of 6.6 may be heated at 100°C for 24 hours without coagulation; it can also withstand heating at 140°C for up to 20–25 minutes. In contrast, the whey proteins are denatured—meaning they undergo dramatic irreversible changes in their structure and properties—on heating at 90°C for 10 minutes, which can lead to their coagulation or gelation. The remarkable heat stability of the caseins (i.e., the ability of milk to be heated to high temperatures without coagulating), due to their lack of extensive secondary and tertiary structures, permits the production of heat-sterilized dairy products with only minor physical changes.

The two protein families also differ in their amino acid composition. The caseins contain high levels of the amino acid proline (17% of all amino acids in β-casein), which disrupts the formation of typical complex structures found in other proteins. They also contain phosphate groups (comprising 1 phosphorus and 4 oxygen atoms, and chemically written as $PO_4^-$, the minus sign indicating a negative electrical charge) attached to the amino acid serine, while the whey proteins do not.

These phosphate groups have major significance for the properties of the caseins, e.g., molecular charge and related properties, such as hydration (their ability to interact with water), solubility, heat stability, and metal-binding ability, which all affect their functional and nutritional properties. Calcium ions bear a double positive charge ($Ca^{2+}$), and the charge being opposite to that of the phosphate groups means that $Ca^{2+}$ has a high affinity for phosphate, as opposite charges on chemical species attract (while similar charges repel). Metal-binding by casein (such as the binding of $Ca^{2+}$ by phosphate) is regarded as an important biological function, since it enables a high concentration of calcium phosphate to be carried in milk in a stable form, to supply the requirements of the neonate. Otherwise, calcium phosphate would precipitate in and block the ducts of the mammary gland, leading to atrophy of the gland and perhaps even death of the animal.

The caseins are low in sulfur (0.8%), while the whey proteins are relatively rich (1.7%). The sulfur in casein is present mainly in the amino acid methionine; the principal caseins lack the other sulfur-containing acids, cystine and cysteine. The whey proteins are relatively rich in cysteine, which has major effects on the properties of these proteins and of milk. As mentioned earlier, cysteine residues are important due to their role in the formation of disulfide bonds between two cysteine amino acid residues, which stabilizes the three-dimensional structure of proteins.

The whey proteins are either found as single free molecules or perhaps pairs of loosely bound molecules (called monomers or dimers, respectively), while the caseins exist as large aggregates, known as micelles, containing about 5,000–10,000 casein molecules. The white color of milk is due mainly to the scattering of light by the casein micelles, as the micelles are of the right size to interfere with the passage of light through the milk sufficiently to create the optical illusion of a white color (systems having particles that create color in this way are called colloidal systems. Clouds are a non-food example of a colloidal system: their white appearance is due to water droplets of a certain size affecting the passage of sunlight).

The structure, properties and stability of the casein micelles are of major significance for the technological properties of milk and the production of dairy products and are described later.

Owing to these differences in properties, the casein and whey protein fractions may be recovered separately from skimmed milk (prepared by centrifuging whole, fat-containing, milk) by several methods:

- Rennet coagulation: Casein micelles are destabilized by rennet and coagulate in the presence of calcium. The properties of rennet-coagulated casein are very different from those of acid-precipitated casein. Previously, so-called rennet casein was used in the production of plastics by polymerization with formaldehyde. One application of such casein was buttons production; now it is more commonly used in food applications, e.g., for making cheese analogues used as pizza toppings.
- Membrane filtration: In processes called ultrafiltration (UF) and microfiltration (MF), which will be described in more detail in Chapter 10, all the milk proteins are retained during filtration through small-pore,

semi-permeable membranes and are separated in this way from smaller molecules such as lactose and soluble salts, which pass through the membranes. UF is used widely in the industrial-scale production of high protein products recovered from whey, called whey protein concentrates (WPCs), and, to a lesser extent, in the production of total extracts of all the milk proteins. Both the caseins and whey proteins are permeable using the larger pore membranes of MF but more than 99.9% of bacteria and other large particles are retained; for this reason, MF can be used for the production of extended shelf-life beverage milk (Chapter 5).

- Caseinates: Casein prepared by adjusting the pH to its isoelectric point is insoluble in water or aqueous systems, but may be converted to water-soluble caseinates by dispersion in water and adjusting the pH to ~6.7, usually with the alkali sodium hydroxide (NaOH) to give sodium caseinate, which may be spray-dried or freeze-dried. Potassium, ammonium, or calcium hydroxide may also be used instead of sodium hydroxide to give the corresponding caseinates.

The caseins and whey proteins also differ in their biological function in milk. The caseins are synthesized in the mammary gland and are unique to that organ. Their function is to meet the amino acid requirements of the neonate and to act as carriers of calcium phosphate, a critical function considering the high rate of bone synthesis in the first weeks and months of life. The principal whey proteins are also synthesized in, and are unique to, the mammary gland, but several minor whey proteins in milk are derived directly from blood; most of the whey proteins have a biological function, some of which will be described later.

## Heterogeneity of the milk proteins

All the major proteins in milk exist in many different forms. For example, all the caseins are phosphorylated, with phosphate groups attached to the amino acid serine, which is critical for the binding of calcium, but they differ in the number of phosphate groups bound. There are also inter-species differences in this: human β-casein contains 1–5 phosphate groups per molecule, while the bovine form of the protein contains either 4 or 5.

The genetics of the mammal also results in inter-individual differences in the exact forms of milk proteins present. In 1955, it was discovered that the bovine whey protein β-lactoglobulin exists in two forms (variants or polymorphs) now called A and B; these differ by only two amino acids and represent the products of genes with very small differences in their DNA sequence (which determines the amino acid sequence). Which variant occurs in milk is thus genetically controlled and the phenomenon is therefore called genetic polymorphism. It is now known that all milk proteins exhibit genetic polymorphism and at least 45 genetic variants of the major milk proteins have been detected. Genetic polymorphism has also been detected in sheep, goat, buffalo, pig, and horse milks and probably occurs in milks of all species of mammals.

Another source of difference between molecules of some milk proteins is the degree of glycosylation—the attachment of sugar residues to a protein—which has a major influence on the solubility of a protein due to the affinity of these sugar groups for water. The only glycosylated casein in milk is κ-casein, and the sugars are galactose and two less common sugars, *N*-acetylgalactosamine and *N*-acetylneuraminic (sialic) acid, which occur as tri- or tetra-saccharides. Up to four such sugars can be attached to each molecule of κ-casein.

The final source of variability in milk proteins is enzymatic breakdown. Milk contains several indigenous proteinases (enzymes that hydrolyze proteins), the principal one being plasmin, a proteinase from blood that plays a key role in the breaking down of blood clots. As mentioned above, β-casein is hydrolyzed rapidly by plasmin, resulting γ-caseins and proteose peptones. The γ-caseins, which typically represent ~3% of total casein in milk, are produced both in the mammary gland and during the storage of milk and occur in all dairy products. Extensive action of plasmin on the caseins (which may occur during mastitis or in late lactation, when levels of the enzyme in milk are elevated) may result in a reduced yield of dairy products or altered properties, e.g., rennet coagulability. This is one way that variations in the quality of milk impact on its processing properties.

## Molecular properties of the milk proteins

The six principal milk proteins are relatively small molecules compared to other proteins, a feature that contributes to their stability. Whey proteins are highly structured, but the four caseins lack stable secondary structures and are called rheomorphic proteins, meaning they are very flexible molecules. Their open, flexible structure renders them very susceptible to digestion by proteolysis, which makes them more readily available as source of amino acids. In contrast, the whey proteins, especially β-Lg, are quite resistant to proteolysis by enzymes during digestion and processing. Since most of the whey proteins have a function other than nutritional, e.g., binding and transporting other molecules such as metals and vitamins, this resistance to proteolysis is important. When consumed, caseins form a clot in the stomach and gradually release their amino acids, while whey proteins, being simpler and more mobile, leave the stomach more rapidly—for this reason, whey proteins are sometimes referred to as being "fast" proteins.

The caseins are often considered hydrophobic (water-hating) proteins but, with the exception of β-casein, they are not exceptionally so. The hydrophobic and hydrophilic parts of the casein molecules are not distributed uniformly, giving them a structure in which some regions are hydrophilic and some regions are hydrophobic. This type of protein structure is called amphiphatic or amphiphilic. Because of this feature, along with their open flexible structure, the caseins are very stable when located at interfaces between water-rich and water-free environments (e.g. between water and oil in emulsions or water and air in foams). This surface-active property makes casein the functional protein ingredient of choice for many food applications, especially emulsions and foams.

Owing to their hydrophobic sequences, the caseins have a propensity to yield bitter hydrolysates, as many peptides that contain hydrophobic amino acids react with bitter taste buds on the tongue. The bitterness of peptides derived from casein is an important flavor element in cheese, where it is toned down or mitigated by other flavor notes but, in hydrolyzed pure casein, bitterness may be a limiting factor in food applications.

The unusual properties of the caseins are reflected in many of their properties during dairy processing. For example, due to their open structure, the caseins form highly viscous solutions, which may adversely affect their processing properties, e.g., it is not possible to spray-dry highly concentrated (>20% protein) casein solutions. The lack of complex structures, on the other hand, renders the caseins very heat stable, which makes it possible to produce heat-sterilized dairy products with very little change in physical appearance; most other major food systems undergo major physical and sensory changes following severe heating (e.g., the coagulation of eggs on cooking).

The caseins have a very strong tendency to associate, due mainly to hydrophobic bonding between regions of the proteins that repel water. Even in sodium caseinate, the most soluble form of casein, the caseins form quite large aggregates of 12–25 molecules. This tendency to associate is important for the formation and stabilization of casein micelles. In contrast, the whey proteins are dispersed as individual molecules in solution.

As mentioned earlier, the caseins contain a high number of phosphate groups, which bind metal ions, and occur in clusters along the molecular structure of the caseins; the main metal bound by them is calcium. Several casein molecules can bind to a calcium ion, which means that these ions can cross-link several casein molecules, causing them to form complexes. For this reason, these caseins are insoluble at even relatively low calcium concentrations at temperatures above 20°C, because the calcium causes large unstable complexes of multiple casein molecules to forms. Since bovine milk contains about 6 times the maximum level of calcium that caseins can bind, it might then be expected that the caseins would precipitate in milk. However, κ-casein, which contains only one phosphate group, binds calcium weakly and is soluble at the calcium concentrations found in dairy products. When mixed with the other, more calcium-sensitive caseins, κ-casein can stabilize ~10 times its mass of these caseins through the formation of casein micelles, as discussed below.

These micelles act as carriers of inorganic elements, especially calcium and phosphorus, but also magnesium and zinc, and are, therefore, important from a nutritional viewpoint. Through the formation of micelles, it is possible to solubilize much higher levels of calcium and phosphorus in milk than would otherwise be possible. This is highly significant nutritionally, since the primary consumers of milk (neonates) are at their most rapidly growing stage of life and such minerals are essential for bone and teeth formation.

## Casein micelles

It has been known since the 19th century that the casein in milk is found in large particles, called casein micelles. The stability of these micelles is critical for many of the technologically important properties of milk, and consequently has been the focus of much research, especially during the past 60 years.[2] Although there remains some disagreement on fine points of the detailed structure of the casein micelle, there is general agreement on their overall structure and on their properties.

Examining milk under powerful electron microscopes shows that casein micelles are spheres with a diameter of 50–500 nanometers (nm, average ~120 nm, where 1 nm is one millionth smaller than a millimeter). They have been shown to have a mass ranging from 1 million to 3 billion ($10^6$–$3 \times 10^9$) Daltons (Da), with an average of ~$10^8$ Da (where 1 Da is the atomic weight of a hydrogen atom; individual casein molecules have masses of about 20,000 Da). The solids portion of the micelles is ~94% protein and 6% low-molecular mass compounds, mainly calcium phosphate, referred to as colloidal calcium phosphate (CCP). The micelles bind ~2 grams of water per gram of protein.

Casein micelles scatter light that passes through milk. This property gives milk its white color, which is lost if the micelle structure is disrupted. They are quite stable when subjected to the principal processes used to produce milk products, such as heat and commercial homogenization. As mentioned above, calcium is present in milk at a level far higher than that possible for individual caseins, demonstrating the critical effect of micelle formation on the ability of milk to contain high levels of calcium.

Some proteinases, generically called rennets, catalyze a very specific hydrolysis of κ-casein, as a result of which the casein coagulates in the presence of calcium. This is the key step in the manufacture of most cheese varieties (see Chapter 6).

The micelles are also destabilized by freezing due to a decrease in pH and an increase in concentration of calcium ions in the unfrozen phase of milk; concentrated milk is particularly susceptible to destabilization on freezing. Casein that has been destabilized by freezing can be dispersed by warming the thawed milk to approximately 55°C, resulting in particles with micelle-like properties.

Based on these observations and properties, there has been speculation since the beginning of the 20th century on how casein micelles are stabilized. But no significant progress was possible until the isolation and characterization of κ-casein in 1956. The first attempt to describe the structure of the casein micelle was made shortly after that and since then, numerous models have been proposed and refined, often leading to robust scientific argument and debate on the merits and limitations of specific models. Any micelle model must explain several features:

- κ-casein, which represents ~12% of total casein, must be located so that it stabilizes the calcium-sensitive $\alpha_{s1}$–, $\alpha_{s2}$–, and β-caseins;
- rennet must be able to rapidly and specifically hydrolyze most of the κ-casein;
- when heated in the presence of whey proteins, κ-casein and the whey protein called β-lactoglobulin must be able interact to form complexes that modify the stability of the micelles; and
- at low temperatures, some casein, especially β-casein, must dissociate from the micelles.

The arrangement that best explains these features has a surface layer of κ-casein surrounding the calcium-sensitive caseins, analogous to a lipid emulsion in which the triglycerides are surrounded by a layer of emulsifier. Therefore, all the major models of the casein micelle propose a surface location of κ-casein.

When the CCP is removed, the micelles disintegrate into smaller particles, suggesting that the casein molecules are held together in the micelles by CCP. For many years, there was strong support for the view that the micelles are composed of smaller entities called sub-micelles linked together by CCP, giving the micelles an open, porous structure and that the micelle disintegrates into these sub-micelles when CCP is removed. Much of the evidence for a submicellar structure relied on electron microscope studies that appeared to show a raspberry-like structure, which was interpreted as indicating sub-micelles. However, this model is no longer widely supported.

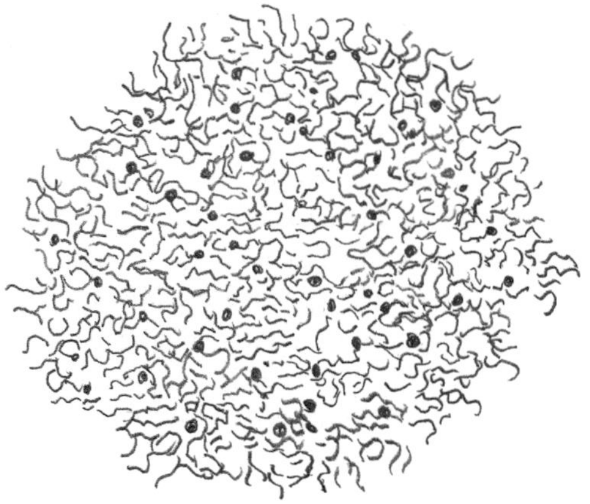

**Figure 2.2** Schematic structure of a casein micelle, showing thousands of strands of individual casein molecules entangled in a roughly spherical structure, within which are found crystals of calcium phosphate (dark circles) (image thanks to Anne Cahalane).

It is now thought that the micelles have a less ordered internal structure and have more of a "tangled ball of wool" nature (Figure 2.2). Various versions of this model have been proposed, and the exact structure remains a lively topic of argument among dairy scientists.

## Whey proteins

About 20% of the total protein of bovine milk is whey (serum) protein. The total whey protein fraction is prepared by any of the methods described above for recovering casein, e.g., they are the proteins that are soluble at pH 4.6 or after rennet-induced coagulation of the caseins.

On a commercial scale, whey protein-rich products may be prepared by a number of methods, in particular membrane filtration of the whey by-product of cheese or rennet casein production to remove varying amounts of lactose, and spray-drying to produce whey protein concentrates (30%–85% protein). Other technologies can yield protein preparations containing ~95% protein, called whey protein isolate. High heat treatment of whey can yield a product that contains all the whey proteins, but has low solubility and poor functionality (this is essentially the principle of producing ricotta cheese).

All these methods yield products that contain a mixture of whey proteins. There is also considerable interest in the production of the major and some of the minor whey proteins on a commercial scale for nutritional, nutraceutical (compounds with beneficial biological functions when present in or added to a food product), or functional applications. For example, a protein called lactoferrin and an enzyme called lysozyme are valuable as antimicrobial proteins that can be recovered from whey and used in applications such as antimicrobial toothpaste.

## β-Lactoglobulin

β-Lactoglobulin (β-Lg) represents ~50% of the whey proteins, equivalent to ~12% of the total protein in bovine milk. It is a typical globular protein with a complex 3-dimensional structure and is the principal whey protein in the milk of cattle, buffalo, sheep, and goat. Initially, it was considered that β-Lg occurs only in the milk of ruminants, but it is now known that a similar protein occurs in the milk of non-ruminants, including the pig, horse, and donkey. However, β-Lg does not occur in human milk, in which α-lactalbumin is the principal whey protein; this absence is particularly noteworthy in terms of infant formulae prepared from bovine milk-derived ingredients, especially as some infants are allergic to β-Lg.

Bovine β-Lg consists of 162 amino acids per molecule and exists as 10 genetic variants (labelled A–J), of which A and B are by far the most common. It contains 5 cysteine residues—the amino acid that forms strong disulfide bonds. In β-Lg, four of the cysteine residues form two intramolecular disulfide bonds (i.e. bonds between different regions of the same β-Lg molecule, which cause the molecule to fold into a complex three-dimensional shape), and one is free (not in a disulfide bond), a relatively rare occurrence in protein structures. On heating milk or whey to a temperature >70°C, the β-Lg molecule unfolds, the free cysteine residue is exposed, and it can then react with a cysteine residue in a different molecule to form a new disulfide bond. Such heat-induced changes are not reversible and result in a new protein structure; this process is referred to as denaturation.

The free (unbound) cysteine residue exposed on denaturation of β-Lg can react with κ-casein with the formation of a new disulfide bond between the two proteins, which significantly affects the rennet coagulability and stability of milk to heat. The exposed sulfhydryl (-SH) group of the free cysteine residue is also responsible for the cooked flavor of heated milk. Because of its ability to form disulfide-linked polymers, β-Lg has very good ability to form strong gels on heating, which is a very valuable property for many applications of whey-derived ingredients, as it can build structure and texture by thickening foods or form semi-solid structures (see Chapter 10).

β-Lg is a highly structured protein with a compact globular structure. It is therefore resistant to proteolysis in its native state, which suggests that its primary function is not just nutritional in terms of supplying amino acids. Two biological roles have been suggested:

- It binds retinol (vitamin A) in a "pocket" within the structure, protects it against oxidation, and transports it through the stomach to the small intestine where the retinol is transferred to a retinol-binding protein, which has a similar structure to β-Lg.
- Through its ability to bind fatty acids, β-Lg stimulates lipase activity, which may be its physiological function.

β-Lg is the most allergenic protein in bovine milk for human infants, and there is interest in producing whey protein products free of β-Lg for use in infant formulae, e.g., through genetic modification of cows. A cow, named Daisy, that produced milk free of β-Lg was reported in New Zealand in 2012.

## α-Lactalbumin

About 20% of the protein of bovine whey (3.5% of total milk protein) is α-lactalbumin (α-La), and it is the principal whey protein in human milk. It is a relatively small protein (the smallest major milk protein) containing 123 amino acid residues and has been well characterized. α-La is a compact, highly structured globular protein that contains four intra-molecular disulfide bonds per molecule but no cysteine amino acids or bound phosphate or carbohydrate groups. For reasons that are not clear, the primary structure of α-La is similar to that of the enzyme lysozyme (the function of which is antimicrobial, as it attacks a key molecule in bacterial cell walls).

α-La has been isolated from the milk of many species, and there are minor inter-species differences in the composition and properties of this protein. As mentioned earlier, α-La is a component of the enzyme system that catalyzes the final step in the biosynthesis of lactose, and there is a direct correlation between the concentrations of α-La and lactose in milk. The milk of some marine mammals contains little or no α-La.

It is a metalloprotein which can bind one calcium ion per molecule in a "pocket" within its structure. The calcium-containing form of the protein is the most heat stable of the principal whey proteins. When the pH is reduced to less than 5, the protein loses its ability to bind calcium. As the metal-free protein is quite heat sensitive, it does not return to its original form on cooling; this property has been exploited to isolate α-La from whey.

α-La is synthesized in the mammary gland, but a little is transferred to blood. The concentration of α-La in blood increases during pregnancy and so is a reliable indicator of mammary gland development and of the potential of an animal for milk production.

## Minor whey proteins

Normal bovine milk contains 0.01%–0.04% of blood serum albumin (BSA; 0.3%–1.0% of total nitrogen), due to leakage from blood in the mammary gland. BSA has no known biological function in milk and, owing to its low concentration, has little effect on the properties of whey-derived protein products.

Mature bovine milk also contains 0.6–1 gram of immunoglobulins (Igs) per liter (~3% of total nitrogen), but colostrum contains around 10% (w/v) Ig, the level of which decreases rapidly post-partum. The Igs are antibodies, which are very important in immune responses to infection. There are several types of Ig, classified by letters, each of which has different specific biological functions. IgG1 is the principal Ig in bovine, goats' and sheep's milk, with lesser amounts of IgG2, IgA, and IgM; IgA is the principal Ig in human milk. Cattle, sheep, and goats do not transfer Ig to the fetus in utero; the neonate is born without antibodies in its blood and therefore is susceptible to bacterial infection with a high risk of mortality. The young of these species can absorb Ig from the intestine for several days after birth and thereby acquire passive immunity from milk until they synthesize their own Ig. In contrast, human mothers transfer immunoglobulins in utero and babies are born with a broad spectrum of antibodies. Although the human baby cannot absorb Ig from the intestine, the ingestion of colostrum is still very important because its Igs prevent intestinal infection.

The modern dairy cow produces colostrum far in excess of the requirements of its calf; surplus colostrum is exploited in some countries for the recovery of immunoglobulins and other nutraceuticals. There is considerable interest in hyper-immunizing cows against certain human pathogens for the production of antibody-rich milk for human consumption, especially by infants. The Ig could be isolated from such milk and presented as a "pharmaceutical" or consumed directly in the milk.

Milk contains several other proteins at very low or trace levels, many of which are biologically active. Some are regarded as highly significant and have attracted considerable attention as so-called nutraceuticals. When ways of increasing the value of milk proteins are discussed, the focus is usually on these minor proteins, but they are, in fact, of little economic value to the overall dairy industry. While found mainly in the whey, some are also located in the MFGM.

Lactoferrin (Lf), an iron-binding glycoprotein (i.e. a protein that has sugars as part of its structure, like κ-casein), is present in several body fluids, including saliva, tears, sweat, and semen and has several potential biological functions. It improves the bioavailability of iron, is bacteriostatic (stopping bacterial growth because the iron it binds is unavailable to intestinal bacteria), and has antioxidant, antiviral, anti-inflammatory, immunomodulatory, and anticarcinogenic activities. Human milk contains a very high level of Lf (up to 20% of total protein) and there is interest in fortifying bovine milk-based infant formulae with it. As Lf is positively charged at the normal pH of milk, while most milk proteins are negatively charged, it can be isolated on an industrial scale from whey by adsorption on negatively charged beads in a process called ion exchange. Digestion of Lf by pepsin yields peptides called lactoferricins, which are even more bacteriostatic than Lf.

Milk also contains proteins that bind vitamins: retinol (vitamin A), biotin, folic acid and cobalamin (vitamin $B_{12}$). These proteins probably improve the absorption of vitamins from the intestine or act as antibacterial agents by rendering vitamins unavailable to bacteria. The vitamin-binding activity of these proteins is reduced or destroyed on heating at a temperature somewhat higher than those used for commercial pasteurization (72–74°C for 15–30 seconds).

Some other minor, biologically active proteins in bovine milk include: osteopontin, one of the major non-collagenous proteins in bone; kininogens, which have an important role in blood coagulation, muscle contraction, and hypertension; and angiogenins, which induce the growth of new blood vessels. Milk also contains many peptide hormones, including epidermal growth factor, insulin, insulin-like growth factors 1 and 2, human growth factors, and platelet-derived growth factor. It is not clear whether these play a role in the development of the neonate or in the development and functioning of the mammary gland, or both.

All milk proteins contain sequences that correspond to peptides that have been demonstrated to have biological or physiological activities and that may be released by proteolysis, such as during digestion. The best studied are phosphopeptides that aid calcium absorption, peptides that might impact on hypertension and blood pressure, platelet-modifying peptides, opioid peptides that might have mood-altering properties, peptides that influence satiety and thus appetite, immunomodulating peptides, and antimicrobial peptides. The presence of these "encrypted" peptides suggests that milk proteins have a function other than simply supplying amino acids and that, when consumed, these peptides are released and travel to different sites in the body where they can exert beneficial health effects.

For this reason, many dairy companies have, in recent years, studied ways to produce these peptides, for example by digesting milk proteins with enzymes in vitro, and then recovering and purifying the peptides before adding them to dairy products at higher levels than normally found, giving products with enhanced biological functionality, and hence higher value. In addition, the production of many fermented dairy products, such as cheese and fermented milks, involves enzymatic hydrolysis of milk proteins, especially by microbial enzymes, and so the health benefits of consuming such products may be linked to the presence of physiologically active peptides.

## Non-protein nitrogen

The non-protein nitrogen (NPN) fraction of milk is defined as those nitrogenous compounds that remain soluble in the presence of 12% trichloroacetic acid (TCA), which precipitates most proteins and peptides. NPN represents ~5% of total nitrogen in milk; its principal components are urea, creatine, uric acid, and amino acids. The urea concentration in milk varies considerably and has a significant effect on heat stability and some other processing properties of milk.

## Indigenous milk enzymes

Enzymes are proteins that act on other molecules to change or modify them; for this reason, they are called biological catalysts. Examples of enzymes already mentioned include rennet, used to coagulate milk for cheese-making; milk lipoprotein lipase, which breaks down milk lipids and causes rancidity; and plasmin, which breaks down milk proteins. Many of the reactions catalyzed by such enzymes involve splitting bonds (such as those between amino acids in proteins) and in the process adding water molecules; for this reason these are known as hydrolysis reactions.

Milk contains about 70 enzymes. These generally either originate from the milk secretory cells or from the blood, which may reflect their function and whether they are specific to milk. Enzymes produced by the mammal themselves are referred to as indigenous enzymes, to distinguish them from those that originate from contaminating bacteria (endogenous enzymes) or are added to milk to make products (exogenous enzymes such as rennet). Many indigenous enzymes are concentrated in the MFGM or in the milk serum, but plasmin and LPL are associated with the casein micelles.

The indigenous enzymes are significant for several reasons:

* Some of them cause deterioration of product quality (e.g. plasmin and lipoprotein lipase, through the breakdown of milk proteins and lipids, respectively).
* Some are bactericidal agents (lactoperoxidase and lysozyme).
* Some can be used as indices of the thermal history of milk (e.g. alkaline phosphatase is inactivated by pasteurization and so is commonly used to determine whether this has occurred, see Chapter 5) or as indices of mastitic infection (the activities of some enzymes such as catalase and acid phosphatase increase during mastitis and so can be used as veterinary diagnostics of animal health).

The concentration/activity of indigenous enzymes in milk shows greater interspecies differences than any other constituent, e.g., 3,000 times more lysozyme is present in equine and human milks than in bovine milk, while human milk contains a lipase stimulated by bile salts, though the milk of most species lacks this enzyme.

## Milk salts

When milk is heated at 500°C for around 5 hours, a grey or off-white ash residue, derived mainly from the inorganic salts of milk, remains. As with other constituents, the concentration of ash in milk varies widely between species and ranges from 0.2% to 1.2% (human milk contains 0.2%, while bovine milk contains 0.7%–0.8%). The elements in the ash are changed on heating from their original forms to derivatives called oxides or carbonates, and the ash contains phosphorus and sulfur derived from caseins, lipids, and sugars. Some organic salts, the most important of which is citrate, are oxidized and lost during ashing; some volatile metals, e.g., sodium, are also partially lost. Thus, ash does not accurately represent all the salt present in milk. However, the principal inorganic and organic ions in milk can be determined directly by other analytical methods.

There is considerable variability in the concentrations of the principal elements in bovine milk, especially in milk from cows in very early or late lactation or suffering from mastitis. Milk also contains 20–25 elements at very low or trace levels, which are important from a nutritional viewpoint; many, such as zinc, iron, copper, and magnesium, are present in enzymes, and some are concentrated in the MFGM. Some micro-elements, such as iron and copper, are very potent lipid pro-oxidants, promoting oxidation and potentially leading to off-flavor formation. Although the salts are relatively minor constituents of milk, they are critically important for many of its technological properties.

The solubility and ionization status (i.e., their electrical charge) of many of the principal mineral species are interrelated, especially calcium, phosphate, and citrate. These relationships have major effects on the stability of the casein system and consequently on the processing properties of milk. The behavior of various species in milk can be modified by adding certain salts to it, e.g., the availability of calcium is reduced by adding salts that bind to it, such as phosphate or citrate, critical in the manufacture of processed cheese (see Chapter 6), while addition of $CaCl_2$ affects the distribution of calcium and phosphate and the pH of milk and is often used to improve the cheese-making properties of milk.

## Vitamins

Milk contains all the vitamins in sufficient quantities to enable normal maintenance and growth of the neonate. Bovine milk is a very significant source of vitamins, especially biotin ($B_7$) and riboflavin ($B_2$), in the human diet.[3]

In addition to their nutritional significance, some vitamins are also significant for other reasons. For example, vitamin A (retinol) and carotenoids are responsible for the yellow-orange color of fat-containing products made from bovine milk; grass-fed cattle have high levels of carotenoids in their milk fat and butter from countries where grass-feeding is extensive (e.g. Ireland and New Zealand) is markedly more yellow than that from countries where this is less common. Vitamin E (tocopherol) is a potent antioxidant, while vitamin C (ascorbic acid) can act as an antioxidant or pro-oxidant, depending on its concentration. Vitamin B2 (riboflavin), which is greenish-yellow, is responsible for the color of whey and ultrafiltration permeate. Riboflavin co-crystallizes with lactose and is responsible for its yellowish color, which may be removed by recrystallization or bleached by oxidation, and acts as a photocatalyst in the development of so-called light-oxidized flavor in milk, due to the oxidation of the amino acid, methionine.

In summary, milk is a very complex fluid. It contains several hundred molecular species, mostly at trace levels. Some of the micro-constituents are derived from blood or mammary tissue but most of the macro-constituents are synthesized in the mammary gland and are milk specific. The constituents of milk may be in true aqueous solution (e.g. lactose and most inorganic salts) or as a colloidal solution (proteins, which may be present as individual molecules or as large aggregates of several thousand molecules, as in the casein micelles), or as an emulsion (lipids). The macro-constituents can be isolated readily and are used widely as food ingredients, as will be explored in later chapters of this book. The natural function of milk is to supply the neonate with its complete nutritional requirements for a period (sometimes several months) after birth, and it also provides many physiologically important molecules, including carrier proteins, protective proteins, minerals, and vitamins.

# 3
# The Microbiology of Milk

## Overview of microorganisms

If raw milk is held at room temperature (15–25°C), it will quickly sour and thicken due to the growth and consequent acid production by microorganisms, particularly bacteria, which contaminate the milk during milking.
Microorganisms are small, living cells most of which can only be seen through a microscope. They include bacteria, yeast, molds, protozoa, algae, and viruses. Yeast and molds together are called fungi and some of them, e.g., mushrooms, are visible to the naked eye. Only the first three groups of microorganisms are important in milk and milk products; representatives of each of them are significant in the making and flavor development (called ripening) of cheese (see Chapter 6).

For example, the bacterium, *Lactococcus lactis*, is involved in acid production during the manufacture of many cheeses and its enzymes are also important in ripening, while the yeasts *Debaryomyces hansenii* and *Geotrichum candidum* and the mold *Penicillium camemberti* are involved in the ripening of washed-rind (e.g. Tilsit) or mold-ripened cheeses (e.g. Brie and Camembert). Generally, bacteria are the more important of the three groups of microorganisms and grow faster than yeast or molds in milk and milk products.

Viruses are also considered to be microorganisms. They are even smaller than bacteria, and an electron microscope is needed to see them. They basically comprise nucleic acid (deoxyribonucleic acid, DNA or ribosomal nucleic acid, RNA) in a protein coat and are not "living" cells in the sense that they cannot multiply by themselves (unlike the other microorganisms, which can, as discussed below) but require a bacterial, animal, or plant cell within which to multiply. Viruses that multiply within bacteria are called bacteriophages and are important inhibitors of acid production in cheesemaking (see Chapter 4).

Bacteria are tiny, single-celled organisms. They were discovered by the Dutch scientist, Antonie van Leeuwenhoek in 1676, in a drop of water he examined using a crude microscope he had made himself. Individual bacterial cells can only be seen at a magnification of 400–1,000 times greater than can be seen by the human eye, except for *Epulopiscium fishelsoni*, a rod-shaped bacterium found in the intestine of the surgeon fish, which is typically 200–600 μm long and can be seen without the help of a microscope. However, masses of microorganisms can

*From Farm to Table*. Alan Kelly, Patrick Fox, and Tim Cogan, Oxford University Press.
 DOI: 10.1093/9780197581025.003.0003

be seen with the naked eye, e.g., algal blooms in water or the surface of a washed-rind cheese, which may contain in excess of 10 billion cells of bacteria and yeasts per $cm^2$.

Bacteria vary in shape and can be spherical (coccal), rod, spiral, corkscrew, or indeed square-shaped. The common shapes are rods or cocci, examples of both of which are used in the manufacture of yoghurt (see Chapter 7). Cocci can exist as chains, in the genera *Lactococcus*, *Streptococcus*, and *Leuconostoc* (these divide in one plane to form simple, almost one-dimensional structures), or in complex clumps, in the genera *Staphylococcus*, *Micrococcus*, and *Macrococcus*. Yeasts are like bacteria, in being unicellular. Oval or cylindrical in shape, they are larger than bacteria, but a microscope is still required to visualize them individually. In contrast, molds are multicellular, very irregularly shaped cells.

Microorganisms are found everywhere, in water, food, particularly fermented foods like cheese, soil, human and animal intestines, and skin and plant surfaces. Up to 100 trillion bacteria (i.e., $10^{14}$ cells), are found in the human intestine and upward of 1 trillion (i.e., $10^{12}$) in 1 kg of cheese.

## The structure of cells

The inside of a bacterial cell is filled with a fluid, called the cytoplasm, that contains the genetic material, DNA, in a single large macromolecule called the chromosome (described below), and a complicated mixture of various components, including water, other macromolecules, e.g., various proteins and RNA, small organic molecules (mainly the components of the macromolecules and vitamins), ribosomes (where proteins are synthesized), and inorganic ions. Some of these, e.g., the DNA and the proteins, are synthesized within the cell during growth and others, e.g., the inorganic ions, are transported into the cell during growth. Other compounds, e.g., sugars, are also transported into the cell, where they are metabolized to obtain energy for growth.

The cell is surrounded by a membrane, called the cytoplasmic membrane, which is about 0.8 nanometers (nm) thick (a nanometer is 1 billionth, $10^{-9}$, of a meter). The membrane is mainly made of molecules called phospholipids, and it is through this membrane that nutrients and waste products of the cell pass. Outside of the cytoplasmic membrane is another layer, called the peptidoglycan layer. This contains a number of specific sugars and amino acids, depending on the bacterium in question, and gives the cell its rigidity.

Some Gram negative bacteria (see below), have an additional layer outside the peptidoglycan, comprising lipids and sugars in a complex called a lipopolysaccharide. Some rod-shaped bacteria also have external "tails," called flagella, attached to the outer part of the cell wall. These are made of protein and allow the bacteria to move and hence show motility when looked at through a microscope. They are called polar flagella if they are found only at one end of the bacterium and peritrichous flagella if they are found emanating from all over the cell. Special stains are needed to see flagella under the microscope, but unstained bacteria, if they have flagella, will appear motile. If they lack flagella, they show a gentle back and forward movement, called Brownian movement, due to buffeting by the random motion of molecules around the cells.

Bacteria in two genera, *Bacillus* and *Clostridium*, also contain small organelles within the cells, called spores. Spores are very heat resistant (they can resist boiling for several hours), and it is because of them that canned foods are sterilized at very high temperatures during production. Spores may be located at the end of the cell (terminal), toward the end (subterminal), or in its center. *Clostridium botulinum*, as well as containing spores, produces a very potent neurotoxin. Its potential presence in the raw material is the main reason for the high heat treatments given to neutral (in terms of pH) or near-neutral canned foods to preserve them (acidic conditions weaken the bacteria), as will be discussed in Chapter 5.

Staining bacteria (and yeast) with brightly colored stains and then looking at them under the microscope is an easy way to determine their shape and size and indeed the number present. Probably the most important stain in microbiology is that developed by Hans Christian Gram in 1884. Consisting of two stains, blue crystal violet and red safranin, it divides bacteria into two groups: Gram-positive, which retain the crystal violet stain and appear blue when examined microscopically, and Gram-negative, which appear as red cells under the microscope.

Bacteria, yeast, and molds have different ways of multiplying. Bacteria multiply by simple division into two identical cells, in a process known as binary fission. Each of the progeny cells contains a copy of the original single chromosome, which is composed of double-stranded DNA. Yeast multiply by budding: a new cell forms a small outgrowth of the old cell and eventually breaks away to form the new cell. Molds multiply by elongation of specialized structures called hyphae.

## Culture and counting of microorganisms

Because microorganisms are found everywhere, all the equipments used to study them must be sterile. Common methods to ensure this include heating at 121°C in the presence of steam for 10–15 minutes (this is called autoclaving because of the specialized piece of equipment, an autoclave, needed for it), heating in a hot-air oven at 170°C for a couple of hours, or by dipping the piece of equipment in alcohol and setting fire to it (flaming). Media for the growth of microorganisms must also be sterile and must contain all the nutrients necessary for the growth of the organism being studied.

Some bacteria have simple nutritional requirements. *Escherichia coli*, for example, is a common inhabitant of cows' intestines, from where it can contaminate the raw milk during milking. It can grow in a simple, defined medium containing sources of carbon (glucose), nitrogen, sulfur, phosphorus, and small amounts of iron, calcium, and magnesium, adjusted to pH 7.4. Other microorganisms, e.g., *Lc. lactis*, are fastidious and require a carbohydrate, several amino acids, and vitamins to grow. Media for this organism generally include a carbohydrate (most commonly glucose), enzymatically hydrolyzed (broken down) casein, meat and/or yeast extract, a suitable buffer like phosphate, various inorganic ions such as calcium, magnesium, manganese, iron, and potassium, and vitamins. The hydrolyzed casein, meat, and yeast extract act as sources of amino acids and vitamins, and the buffer helps to maintain the pH in a fairly neutral range (pH 5.0–8.0, where pH 7.0 is neutral). Milk is a perfect medium for the growth of bacteria, because it has almost a neutral pH (6.6) and contains all the nutrients that even fastidious bacteria like *Lc. lactis*, require for growth, including protein, a source of amino acids, sugar (lactose), vitamins, various inorganic ions including phosphorus and calcium, and, of course, water.

Agar, a polysaccharide derived from seaweed, has the unusual property of remaining liquid above 45°C and solidifying below that temperature, and is added to many media used to enumerate live (viable) microorganisms in a sample of milk or other food. There are two ways for doing this, the pour plate and the spread plate methods. In both methods, the milk is diluted aseptically 10-, 100-, or 1,000-fold etc., in a suitable, sterile buffer. In the pour plate method, aliquots, generally 1 ml, of each dilution are placed in a petri dish to which is added a suitable sterile, nutrient solution, e.g., Plate Count Agar, which has been sterilized and cooled to 45°C. After it has been mixed to distribute the aliquot, the agar hardens. As well as causing the nutrient solution to solidify, the agar also prevents the bacterial cells from spreading throughout the agar during growth. The petri dish and agar are incubated at 30°C for 48 hours, during which time each cell in the aliquot grows, producing visible colonies containing several million bacteria. The number of colonies on plates containing between 30 and 300 colonies are counted and multiplied by the dilution factor to obtain the number of bacteria/ml of the sample. The spread plate method is very similar except that 0.1 ml of each dilution is spread on the surface of poured and hardened medium, using a sterile bent glass rod, called a hockey stick. Following incubation, the number of colonies are counted and multiplied by 10 (to convert to milliliters) and by the dilution factor to obtain the number of colonies/ml. To isolate pure cultures of microorganisms, a similar method is used: a small amount of the colony is emulsified in sterile diluent and spread with a sterile loop onto the pre-hardened agar and incubated for a few days to obtain isolated colonies.

The absence or presence of oxygen is also important for growth, so sometimes agar plates are incubated anaerobically (excluding oxygen) to determine how the bacteria react to air, or how many anaerobic bacteria (which cannot grow in the presence of oxygen) or facultatively anaerobic bacteria (which can grow in the presence or absence of oxygen) are present in a particular sample. A good example of a facultative organism (which can grow aerobically and anaerobically) is *E. coli*. An organism that grows only in the presence of air is an obligate aerobe (e.g., *Pseudomonas* spp.), while one that grows only in the absence of air is an anaerobe (e.g., *Clostridium* spp.). Microaerophilic bacteria, e.g., *Campylobacter jejuni*, a common cause of food poisoning, grow best in an atmosphere of several gases, e.g., 5% oxygen, 20% carbon dioxide, and 85% nitrogen.

Bacteria can also be counted under a microscope by etching one square centimeter on a glass slide and spreading a known volume of milk (usually 10 microliters) uniformly over the area. After staining with methylene blue and drying the slide, the number of bacteria in several fields are counted, averaged, and multiplied by the microscopic factor to give the number of bacteria/ml. However, this process does not distinguish between dead and live cells.

Yeast and molds are resistant to acid and the medium used to count them is usually acidified to around pH 3.5, through the addition of sterile lactic or tartaric acid, prior to pouring to enumerate these organisms.

The growth of bacterial cells is relatively simple; one cell divides to produce two cells, two cells divide to produce four, four cells divide to produce eight, eight cells divide to produce sixteen, etc. Mathematically, this is an exponential or logarithmic progression, and so growth curves for bacteria involve plotting the logarithm of the number of bacteria or some other estimator of bacterial growth, e.g., percent developed lactic acid (the total amount of lactic acid produced minus the amount present in the uninoculated medium) in the case of starter bacteria, against time (Figure 3.1). Such plots generally show four phases: a lag phase effectively showing no growth, followed by an exponential phase, when rapid growth occurs, a stationary phase, when no growth occurs, and a death phase, when the death of many cells occurs. The doubling or generation time can be calculated from the slope of the exponential phase; for many bacteria this is less than 30 minutes. In milk at 30°C, *Lc. lactis* has a doubling time of about 1 hour at 30°C.

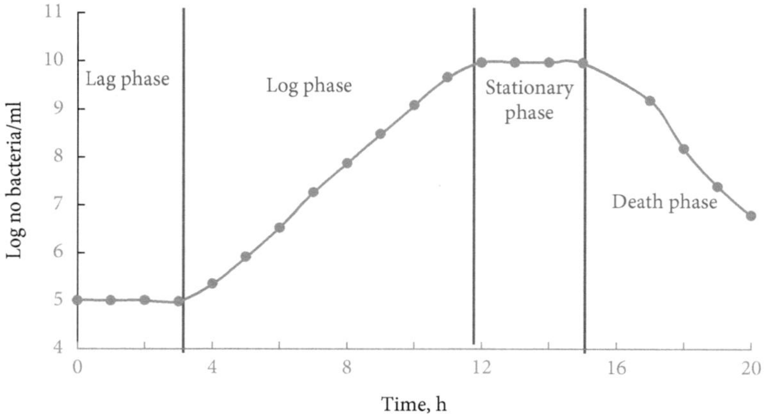


**Figure 3.1** A typical bacterial growth curve where the logarithm of the number of bacteria is plotted as a function of time. The four normal growth phases are outlined.

## Nucleic acids

All cells contain two types of nucleic acid, DNA and RNA. The DNA contains the information (the genetic code) for the synthesis of all the proteins necessary for cell growth and multiplication, while the RNA is involved directly in protein synthesis. The segment of the DNA that encodes a protein is called a gene, and so there are at least as many genes in the DNA sequence as the number of proteins necessary for growth.

The building blocks of DNA are nucleotides, which are composed of one of four nitrogen-containing bases: adenine (A), guanine (G), cytosine (C), and thymine (T). A and G are called purines and have a two-ringed structure; C and T are pyrimidines and have a one-ring structure. Each base is attached individually to a 5-carbon sugar, deoxyribose, which, in turn, is attached to a triphosphate. The base-sugar combination is called a nucleoside, the base-sugar-phosphate is called a nucleotide, and the complete sequence of bases, sugars, and phosphates is called a polynucleotide.

A DNA molecule is made up of two complementary strands of polynucleotides with the sugar-phosphates forming the backbone and the purines and pyrimidines sticking out from it. Bonds called hydrogen bonds form between the A on one strand and the T on the other strand and between G on one strand and C on the other. This means that, whenever an A is found on one strand, a T is found on the other, and whenever a G is found on one strand, a C is found on the other. The pairing of A with T and G with C is energetically the most stable structure, allowing the formation of the famous double helix. The unravelling of the structure of DNA by James Watson and Francis Crick in 1953 was arguably the greatest development in biology in the 20th century and earned them, together with Maurice Wilkins, the Nobel Prize in Medicine in 1962. A schematic of the linear and helical structures of DNA is shown in Figure 3.2. The DNA macromolecule is long and thin but, because it forms a double helix, it is supercoiled in a structure called the chromosome. Each bacterial cell contains only one chromosome, while yeast cells contain 16, and human cells 46.

RNA has a similar structure to that of DNA except that ribose (another 5 C sugar) replaces deoxyribose, and a different pyrimidine base, uracil, replaces thymine. In contrast to DNA, all RNAs are single-stranded, except for those in some viruses, which are double-stranded. There are three types of RNA: messenger RNA (mRNA), transfer RNA (tRNA), and ribosomal RNA (rRNA). All of them are involved in protein synthesis. This is quite complex and will not be discussed further, except to say that the information in the gene sequence is transcribed into mRNA, which is complementary to the DNA sequence. This is, in turn, translated by tRNA into the amino acid sequence of the particular protein being synthesized. This complexity can be simply envisaged as:

DNA → RNA → Protein

## Identifying and naming bacteria

All living organisms, whether animals, plants, or bacteria, are thought to have evolved from a common ancestor a few billion years ago. Every organism, whether a bacterium, an animal, or a plant has two names, a concept developed by the Swede, Carl Linneus, in 1736. The first name is that of the genus (plural genera)—a group of related species—and the second is the name of the species—a group of similar strains. When written both names are usually italicized and the genus is abbreviated to the first one or two letters after its first mention in a text. Generally, the names are derived from Latin or Greek, e.g., *Lactococcus lactis* is a common species found in the starter cultures that are deliberately added to milk to produce lactic acid in the making of cheese. *Lactococcus* is the genus name and is a combination of the Latin word *lac* meaning milk and the Greek word *coccus* meaning berry or a grain, while *lactis* is the species name and is the Latin for "of milk," hence the name evokes a "milk grain" or "berry from milk." Other species are named after a person, e.g., the genus *Listeria* is named after Joseph Lister, a Scottish surgeon who developed aseptic (sterile) surgery, using carbolic acid as a disinfectant. Occasionally, a bacterium is given a third or subspecies name, when

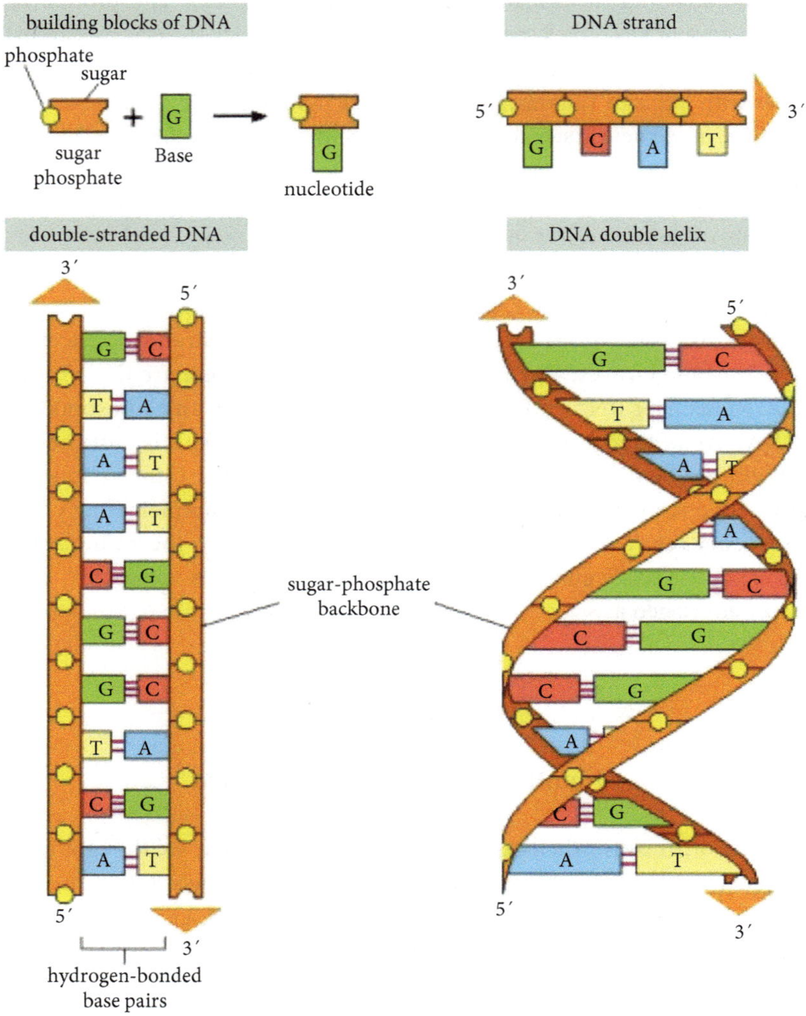


**Figure 3.2** Structure of deoxyribonucleic acid. Both linear and helical structures are shown for clarity. G, C, A, and T are the bases, guanine, cytosine, adenine, and thymine. See text for more details (from Alberts et al. 2002. *Molecular Biology of the Cell.* 4th edn. Garland Science, New York, with permission).

two organisms are very closely related, e.g., *Lc. lactis* subspecies (subsp.) *lactis* and *Lc. lactis* subsp. *cremoris* (the Latin for "of cream").

In 1977, Carl Woese and George Fox divided all organisms, from the smallest (bacteria) to the largest (blue whale) into three major domains—Bacteria,

Archaea, and Eukarya—based on the sequence of the 16S rRNA gene, in the case of the first two domains, and the 18S rRNA gene in the case of the Eukarya. Ribosomes are complex structures composed of protein and RNA and are involved in protein synthesis. These two molecules were selected because much of their nucleotide sequences have changed little over the millennia (the sequences are referred to as being conserved).

Characteristic properties of Eukarya include well-defined nuclei and the presence of mitochondria, the energy-producing factories of cells. All animals and plants are grouped in the Eukarya, as also are yeast and molds. In contrast, bacteria do not have a well-defined nucleus nor mitochondria. The domains are subdivided into various groupings called, in order: Division, Order, Family, Genus, and finally, Species, as introduced in Chapter 1 for classifying mammals. A division contains several orders, an order contains several families, etc. A genus contains several species, but occasionally a species can contain several subspecies, which are a group of organisms that are very closely related genetically. A good example is the starter bacterium, *Lactococcus lactis* subsp. *lactis* and its closely related species *Lactococcus lactis* subsp. *cremoris*. Interestingly, *Lc. lactis* was the first bacterium isolated in pure culture, also by Lister in 1873.

Archaea are also bacteria, but they differ from the true bacteria (Eubacteria) in not containing peptidoglycan in their cell walls and in the structure of their triglycerides. Many archaea are found in extreme environments like salt lakes—these are called halophiles (salt-loving, growing optimally in 12 to 23% NaCl)—hydrothermal vents (one of them has been shown to grow at 110°C), or in anaerobic conditions, where they may produce methane from carbon dioxide and hydrogen. The easiest way to distinguish archaea from bacteria is that archaea thrive in extreme, harsh environments and most bacteria do not.

Before a microorganism can be identified, a pure culture must be obtained by the procedure already described. Traditionally, the first tests done on the purified organism are its shape (using the microscope), its Gram reaction, how it responds to growth under aerobic and anaerobic conditions, and whether it is motile or not. Other important tests include its ability to hemolyze blood (cause blood cells to burst), and the different sugars it uses as energy sources.

Nowadays, if the sophisticated equipment necessary is available, the partial nucleotide sequence of the conserved region of the 16S rRNA gene (bacteria) or 18S rRNA gene (yeast) or the complete sequence of the ~1,500 nucleotides in the gene are determined, and the results compared to sequences in large databases. Such analyses allow researchers to name the organism immediately or determine if it is a new organism, without the need to do all the classical tests. The term microbiome has become common in recent years to refer to the diverse microbial populations of analyzed materials, and advanced molecular biology techniques such as those mentioned have shown that dairy products such as cheese have extremely diverse microbial ecologies, typically containing many more types of microorganisms than those added during manufacture or that grow during ripening.

## Metabolism

Before any compound can be metabolized by a microorganism, it must be transported into the cell—several complex mechanisms are used for this purpose, which will not be discussed further. The energy for the growth of most common bacteria is obtained from the complete oxidation of sugar (generally glucose) to $CO_2$ and water. This is called respiration. Lactic acid bacteria are unable to respire and either partially oxidize the glucose to lactic acid in a series of steps called glycolysis (species that do this are called homofermentative and include lactococci and homofermentative lactobacilli) or partially oxidize the glucose to lactic acid, ethanol, and $CO_2$ (species that do this are called heterofermentative and include leuconostocs and heterofermentative lactobacilli). These pathways are called fermentation, and oxygen is not involved. A similar process occurs in yeast, but ethanol and $CO_2$ are produced instead of lactic acid; the $CO_2$ is responsible for the bubbles in kefir after fermentation by the yeast.

In both fermentation and respiration, adenosine triphosphate (ATP) is the energy compound produced and is the source of energy for cell growth and synthesis of DNA, RNA, proteins, lipids, and other metabolic products. Respiration is much more efficient in producing ATP than fermentation.

## Types of bacteria

Bacteria can be considered as good or bad. The good ones include starter bacteria, which are necessary to ferment milk sugar (lactose) to lactic acid in the making of fermented foods like cheese and yoghurt, while the bad ones include those that cause disease, like *L. monocytogenes*, which causes listeriosis, and *Mycobacterium tuberculosis*, which causes TB.

Bacteria can also be divided into psychrophiles (cold loving), with an optimum growth temperature of 4–15°C; mesophiles (moderate loving), with optimum temperatures of 30–37°C; and thermophiles (heat loving), with an optimum temperature around 50°C. Some microorganisms in hot springs can grow at 110°C, 10 degrees above the boiling point of water!

Psychrotrophs (not to be confused with psychrophiles) are bacteria that can grow at 4–6°C but whose optimum temperature is generally much higher

(15–30°C). Because they can grow at refrigeration temperatures, contamination of raw milk with these bacteria is a particular nuisance.

## Factors affecting microbial growth

Several factors affect the growth of microorganisms, e.g., temperature, the presence of NaCl and other salts like nitrate, pH, heat, and the amount of water present.

Temperatures above and below the optimum slow down the rate of growth of microorganisms. This is the reason why food is normally stored at refrigeration temperatures of ~4°C. As raw milk is an excellent medium for the growth of bacteria, it should be cooled quickly to 4°C after milking to retard the growth of any contaminating bacteria that might be present. However, it must be remembered that psychrotrophic bacteria can grow, albeit slowly, at this temperature.

The salting of food with NaCl is probably the oldest method of food preservation. NaCl reduces the growth rates of microorganisms; if the concentration is sufficiently high, it prevents their growth completely. However, the tolerance of different bacteria to salt varies significantly. For example, *Lc. lactis* subsp. *cremoris*, a common starter culture for Cheddar cheese (see Chapter 6), grows in 2% but not 4% NaCl, while *Lc. lactis* subsp. *lactis* grows in 4% NaCl. In contrast, most strains of *Staphylococcus aureus*, a common cause of food poisoning, can grow in 10% NaCl. Enterococci are another group of lactic acid bacteria that are quite salt tolerant and can grow in 6.5% NaCl, while corynebacteria, commonly found on the surface flora of red smear cheese, can grow in 10%–12% NaCl.

The salt in foods is dissolved in the water contained in the food, so the effective salt concentration is higher. For example, a typical Cheddar cheese containing 35% moisture and 1.8% NaCl has an effective salt concentration of 1.8 x 100/35 or 5.14%. Salted butter may contain 1.5% salt and 16% water; as all the salt is dissolved in the water, the effective salt concentration is 9.4%.

Sodium or potassium nitrate is often added to milk before cheesemaking (and indeed other foods, e.g. pork in the production of bacon), to prevent the germination of spores and subsequent growth of clostridia, particularly *Clostridium tyrobutyricum* and *Cl. butyricum*, which cause late gas formation in cheese and unsightly development of huge eyes, due to production of carbon dioxide and hydrogen. Hydrogen is the main cause of gas formation as, unlike carbon dioxide, it is insoluble in water. The major source of clostridia is silage, which is forbidden to be fed to cows producing milk for manufacture of Emmental cheese in Switzerland and Parmigiano Reggiano cheese in Italy. The actual inhibitor is nitrite ($NO_2$), which is produced from the added nitrate ($NO_3$) through the activity of an enzyme present in milk, called xanthine oxidoreductase. While nitrate is a traditional food additive, and is even found naturally in some foods such as green leafy vegetables, its use in food has been questioned because it can react with amino acids to produce nitrosamines, which have been implicated as the cause of some cancers.

Most bacteria grow best at neutral pH (i.e., pH 7.0) or slightly below. Milk has a pH of 6.6, while Cheddar cheese has a pH of 5.3. Yoghurt has a pH of 4.6., at which growth of bacteria will be significantly reduced.

Heating at a high temperature kills (inactivates) bacteria, mainly due to the denaturation of proteins in the bacterial cell. Some bacteria are more heat resistant than yeast or molds. Pasteurization, using heat to preserve food by exploiting this sensitivity to heat, was first introduced by Louis Pasteur in the 1860s to reduce the spoilage of wine. In the dairy industry, pasteurization of milk at 63°C for 30 minutes in the so-called low temperature long time (LTLT) method or 71.7°C for 15 seconds in the high temperature short time (HTST) method has been shown to inactivate *Mycobacterium tuberculosis* (the bacterium that causes the disease tuberculosis) and all other pathogenic organisms that might be present in milk. This also increases the shelf-life of pasteurized milk compared to raw milk from 1 or 2 days to perhaps 2 weeks or more, if the pasteurized milk is stored at a low temperature. Pasteurization of milk is discussed in detail in Chapter 5.

Water is another key factor influencing bacterial survival and growth. It is not really the amount of water that is the controlling factor but rather the water activity ($a_w$), which is a measure of the availability of water in a system and ranges from 0 to 1, with pure water having a water activity of 1. A value of 0 would indicate that there is no water available to support, for example, chemical reactions or microbial growth. As examples, the $a_w$ of milk is 0.99, that of cheese is around 0.95, while milk powder has an $a_w$ of 0.2. The $a_w$ of cheese decreases during ripening due to two factors:

1. the breakdown of the protein and fat and the consequent uptake of water, making less water available; and
2. the fact that many cheeses, if they are not packaged, lose moisture by evaporation during ripening, especially if the ripening period is long.

Bacteria have a higher minimum $a_w$ (at 0.90) for growth than yeast (0.88), which in turn have a higher minimum $a_w$ than molds (0.8).

No one factor is responsible for controlling the growth of microorganisms in food but all of them act in concert as so called "hurdles" (or barriers to bacterial growth), which synergistically inhibit bacterial growth, as cells cannot easily withstand several simultaneous stresses. The main hurdles in cheese are the low pH, the presence of salt, low ripening temperatures, and relatively low water activity.

## Contamination of milk by microorganisms

Milk within the udder of healthy cows is usually sterile, but small numbers of microorganisms can enter the teat canal through the teat orifice. Some of these, particularly *S. aureus*, but also *Streptococcus agalactiae* and *Campylobacter jejuni*, can grow in the udder and cause mastitis, a bacterial inflammation of the udder. Mastitis is often diagnosed by measuring the increase in the number of somatic cells in milk that occurs because very high numbers of somatic cells (white blood cells) enter the udder (and hence the milk) to fight the infection. Other bacteria, like coliforms, can cause mastitis, but they are a much less common cause. Diseases of the udder are treated by infusing antibiotics, like penicillin or one of its derivatives, into the teat orifice. This can give rise to residues of the antibiotic in the milk in the subsequent four or five milkings, which can inhibit the growth of the starter bacteria in cheese-making. Milk from the teats of cows treated with antibiotics should not be mixed with the rest of the herd milk because all the milk will then be contaminated with the antibiotic, which inhibits the growth of starter cultures in cheese and yoghurt and also has public health implications, as it encourages the development of antibiotic-resistant bacteria.

Key sources of microorganisms in milk include the environment like air, animal bedding, feces, soil, water, mastitis, the cows' teats, and the milking machine and milk storage tank on the farm (Figure 3.3). Of these, the milking machine and its associated pipework and the bulk milk storage tank are the most important sources of contamination, which is why particular attention is always given to ensuring that they are properly cleaned. The final rinse given to equipment is often a dilute solution of a sterilant like sodium hypochlorite. This is unstable and breaks down to chlorine, which acts as the disinfectant, and water. It is not rinsed off the equipment but allowed to drain to the lowest point. Any residual disinfectant is decomposed when it comes in contact with the milk. To prevent spoilage of the milk, it should be cooled to 4°C in a refrigerated bulk tank and kept at this temperature for 2–3 days until it is collected.

During the storage of raw milk in bulk tanks at 4°C, psychrotrophic bacteria, especially Gram-negative *Pseudomonas* spp., can grow to high numbers ($10^6$/ml), but are usually killed by pasteurization. Some of the enzymes they produce, though, are heat-stable and not inactivated by the heat treatment, e.g., lipases (which hydrolyze fat to glycerol and fatty acids) and proteinases (which hydrolyze proteins to amino acids and small peptides). These reactions can cause off-flavors to develop in the milk or other dairy products, especially if they are held at room temperature. Pasteurized milk has a shelf life of 16–22 days, but the entry of a rapidly growing psychrotroph into milk as a post-pasteurization contaminant can reduce the shelf-life significantly, to perhaps 2–4 days.

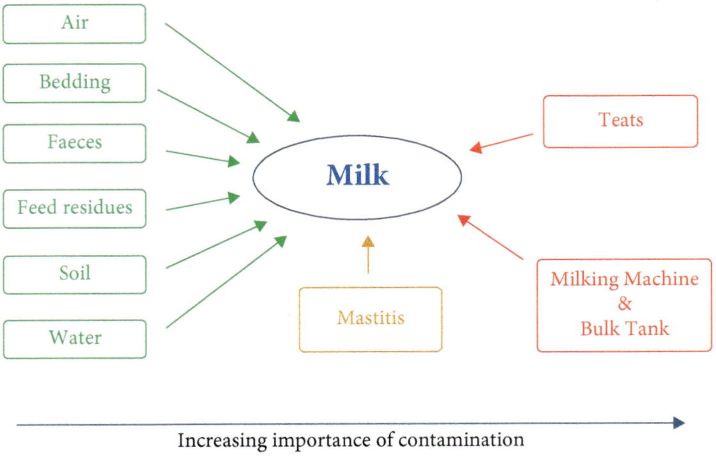


**Figure 3.3** Sources of contamination of raw milk. The most important sources are indicated in red while those of intermediate importance are colored yellow and those of least importance are in green (from Fox et al. 2017. *Fundamentals of Cheese Science*. 2nd edn. Springer, New York, with permission).

There are also microorganisms that can withstand pasteurization. These are called thermoduric bacteria, and include mainly Gram-positive rods, like *Microbacterium* and *Corynebacterium* spp., cocci, like *Micrococcus* spp., and spore-formers. Thermoduric bacteria grow slowly in milk; almost invariably, high numbers of thermoduric bacteria indicate that the milking equipment was not cleaned properly.

Raw milk also contains natural inhibitors of bacterial growth, such as the enzyme lactoperoxidase (LP), which in association with hydrogen peroxide and a minor milk constituent called thiocyanate can prevent the growth of bacteria in milk for 24 hours at 15°C or 6–8 hours at 30°C. This is occasionally practiced in some countries (e.g. in Asia) where maintenance of a cold chain for milk storage is not always possible. LP is not inactivated by conventional pasteurization and inhibits the growth of *B. subtilis*, *Sc agalactiae*, and some strains of *Lc. lactis*, but not *S. aureus*, *E. coli*, or *Ec. faecalis*.

## Milk production regulations

Regarding raw milk, regulations in individual countries or regions like the European Union (EU) specify the health requirements for milk production, hygiene in milk production units, and the microbiological standards and temperature to which raw milk should be cooled. According to EU regulations, raw milk should have a plate count of <100,000 bacteria/ml and a somatic cell (white blood cells) count of <400,000/ml when received at the manufacturing plant and should be cooled quickly to not more than 6°C until it is processed. Raw milk can be kept at a higher temperature only if processing begins within 4 hours of it being delivered to the processing plant. These criteria are very easy to meet in practice, and today, with developments in detergents and sterilizers, it is relatively easy to produce milk with <2,000 bacteria per ml, whereas 40 years ago it was difficult to produce it with <500,000 bacteria/ml.

## Indicator organisms

Coliforms are Gram-negative, motile or non-motile bacteria that produce acid and gas from lactose and include bacteria from the genera, *Escherichia*, *Klebsiella*, *Enterobacter*, *Hafnia*, and *Citrobacter*. Many species in these genera are of intestinal origin, and so the presence of coliforms in milk and water is often used as an indicator of fecal contamination and the potential presence of pathogens like salmonella and shigella, which are also intestinal bacteria but do not ferment lactose and die off faster than coliforms, at least in water. Enterococci are also found in animal and human intestines and are also used as indicators of fecal contamination. Selective media are used to detect them.

## Important pathogens in milk and dairy products

Dairy products can be a cause of food-borne infection of which there are two types: *food poisoning* caused by the production of toxins in the product *before* consumption; and *food infection*, caused by microorganisms growing in the intestines *after* consumption. Symptoms of both food poisoning and food infection usually involve stomach cramps and diarrhea, which sometimes progresses to bloody diarrhea, fever, and vomiting An analysis of US records from the National Outbreak Reporting System run by the Centers for Disease Control for the period 2000–2018 showed that there were 57 outbreaks of food-borne disease associated with the consumption of raw milk, primarily caused by *Campylobacter* spp., followed, in turn, by *E. coli* and *Salmonella* spp.

*Campylobacter* are the most common cause of gastroenteritis worldwide. They are Gram-negative, spiral-shaped bacteria and are microaerophilic, requiring low levels of oxygen to grow. They are unusual organisms in that they neither ferment nor oxidize carbohydrates; instead, they obtain energy from the metabolism of amino acids or citric acid. Their infective dose is low, about 500 cells. The most common source of campylobacter is poultry feces, but they are also found in the feces of many birds and animals.

*E. coli* is a Gram-negative, facultatively anaerobic (able to grow in the presence and absence of oxygen), rod-shaped bacterium commonly found in the intestinal tract of humans and warm-blooded animals. In the past, *E. coli* was assigned to numerous serovars (or varieties), based on their somatic (O) and flagellar (H) antigens (an antigen is a molecule capable of interacting with specific components of the immune system). The somatic antigens are complex lipopolysaccharides associated with the cell walls of the organism, while flagellar antigens are associated with flagella, the protein appendages on some bacteria that allow them to move about. Most strains are harmless, but some are pathogenic, causing diarrhea, urinary tract infection, meningitis, and kidney failure. Those that cause disease are now defined on the basis of specific virulence genes as enteropathogenic *E. coli* (EPEC), enterotoxigenic *E. coli* (ETEC), enterohemorrhagic *E. coli* (EHEC) and Shiga-toxin producing *E. coli* (STEC), depending on how they cause disease.

*E. coli* O157:H7 is both an EHEC and an STEC strain, which produces toxins, known as Shiga toxins because of their similarity to the toxins produced by *Shigella dysenteriae* (an organism that causes dysentery and is closely related to *E. coli*); it causes bloody diarrhea, hemolytic uremic syndrome (HUS), and kidney failure. Raw milk can be a reservoir of *E. coli* O157:H7, and the infective dose is very low, around 1 cell per ml. Other strains of *E. coli* can cause non-bloody diarrhea.

*Salmonella* are also Gram-negative, facultatively anaerobic, rod-shaped bacteria. They are closely related to *E. coli* genetically and have been incriminated as the cause of food poisoning due to the consumption of raw or pasteurized milk, Cheddar cheese, ice cream, and milk powder. *Salmonella* cause typhoid fever and enterocolitis; the symptoms include diarrhea, fever, abdominal pain, and headaches. They are widespread in the environment, but their ultimate source is human and animal feces; chicken meat and eggs are particularly common causes of salmonella. The infective does is very low, around 1–10 cells, and the cause of the infection is a molecule called lipopolysaccharide in the outer membrane of the salmonella cell. Live cells have to be ingested to cause the infection and the symptoms (diarrhea, stomach pains, nausea, and vomiting) begin 6 hours to 3 days later. Recovery without treatment generally occurs within 5 days. Historically, salmonellae have been clinically categorized as invasive (causing typhoid, a serious debilitating disease) or non-invasive, based mainly on the manifestation of the disease. Like *E. coli*, *Salmonella* can be differentiated on the basis of different somatic O and H (flagellar) antigens and more than 2,500 serovars (or serotypes) have been described. Advances in molecular techniques have shown that the genus *Salmonella* comprises only two species, *S. enterica*, which is divided into 6 sub-species, and *S. bongori*.

Listeriosis is caused by *L. monocytogenes*, which is a Gram-positive, facultatively anaerobic, rod-shaped bacterium that causes septicemia and meningitis in young and elderly people, especially if they are immuno-compromised, and spontaneous abortion in pregnant women. Infection of the blood stream, the central nervous system, and the fetus during birth by mothers who show no obvious signs of infection are common. Listeriosis is an atypical foodborne illness because of its viciousness—there is a 20%–30% mortality rate. Soft cheeses, where the surface pH increases during ripening, have been incriminated in several outbreaks of listeriosis. Important symptoms include vomiting and diarrhea. Listeria are salt, pH, and temperature-tolerant, growing in the presence of 10–12% NaCl, at pH values of 4.4. and at −4°C. Such properties make them problematic, particularly in soft cheeses.

The ability of *Salmonella* and *E. coli* to cause disease involves colonization and attachment of the bacterium to intestinal epithelial cells, while listeriosis involves translocation of the bacteria from the stomach to the liver and spleen—the infectious dose is low, probably less than 100 cells (compare this number to the number of bacteria in milk after addition of starter in cheese and fermented milk production, which is $\sim 10^7$ cells/ml of milk).

Other organisms can also cause food-borne disease. *Cronobacter sakazakii*, formerly called *Enterobacter sakazakii*, is a Gram-negative rod that causes meningitis and bacteremia in infants, for which mortality rates range from 20%–50%. It has been detected in powdered infant formula and cheese. There is still a lack of information on its natural habitat, but it has been isolated from diverse environments, including food processing plants and domestic environments.

*S. aureus* is a Gram-positive, facultatively anaerobic coccus that occurs as clusters. Unlike *Listeria*, *Salmonella*, and *Escherichia* spp., food poisoning caused by *S. aureus* is due to heat-stable protein toxins (they can withstand heating to 100°C for 30 minutes), formed in the food before it is eaten, many of which are produced during the exponential (active) phase of growth of the organism. At least 21 different enterotoxins (toxins which cause intestinal problems) are known, and 1 μg of such toxin is sufficient to cause food poisoning; this implies that the infectious dose of staphylococci is quite high, of the order of $10^6$ cells/gram of food. *S. aureus* is the most common cause of mastitis in dairy cows, from where it can contaminate the raw milk; many mastitic strains also produce enterotoxins. This organism is also commonly found on human skin and nasal passages. The symptoms are nausea, vomiting, and abdominal cramps, and occur within 30 minutes to 6 hours after ingestion. *S. aureus* is quite salt-tolerant and can grow in up to 20% NaCl. Staphylococcal food poisoning differs from other causes of food poisoning in that it is the toxin that causes the poisoning rather than the organism itself.

*Clostridium botulinum* is an anaerobic, Gram-positive, spore-forming rod, commonly found in soil and water, and produces one of the deadliest toxins known. It causes botulism in dairy cows, which is on the increase because of the use of plastic-wrapped and non-acidified silage as cattle feed. Botulism in humans can be fatal and between 0.1 and 0.5 μg of toxin is sufficient to cause the disease. It acts by blocking nerve function; the symptoms include difficulty chewing and swallowing and drooping eyelids. Curiously, despite its toxicity, botulinum toxin or botox is used in human medicine to treat twitching of the eye, dry eye, or migraines, and is used cosmetically to reduce wrinkles on the skin.

Bacteria also cause disease in dairy cows: *Mycobacterium tuberculosis*, the cause of TB in humans, also causes this disease in cows; *Brucella abortus* causes contagious abortion in cows (and brucellosis or undulant fever in humans); and *Salmonella* spp. cause salmonellosis. An outbreak of TB in a child in Ireland in 2005 was traced to the consumption of raw milk from the farm on which the child lived.

Cows infected with *Salmonella* and *L. monocytogenes* may shed them in their milk: outbreaks of salmonellosis and listeriosis have been traced to individual cows shedding 200 and 280 cells of *Salmonella* and *L. monocytogenes* per ml, respectively, from individual quarters without any signs of clinical disease or abnormality in their milks.

All of these organisms, except the spores of *Cl. botulinum*, are killed by pasteurization, which is therefore an important method of ensuring the safety of dairy products. However, many cheeses are still made traditionally from raw milk; this is discussed in Chapter 6.

# 4
# Starters for Fermented Dairy Products

The production of lactic acid from lactose is a fundamental step in the manufacture of most fermented dairy products (whey cheeses are an exception) and is carried out by adding small amounts (about 1%) of different species of lactic acid bacteria (LAB) to the milk (for cheese and fermented milk production) or cream (for butter production) and holding the temperature at values that allow the culture to grow. This culture is called a starter because it starts or begins the production of the lactic acid and traditionally is grown in milk. Growth of the starter requires energy. This is obtained through the production of lactic acid from lactose, which, in turn, results in a decrease in the pH of the product being made.

## Some history

In 1873, the main bacterium that causes milk to sour was isolated by Joseph Lister. While Lister named it *Bacterium lactis* ("the bacterium of milk" in Latin), it was renamed *Streptococcus lactis* in 1909 and renamed *Lactococcus lactis* in 1987. This microorganism was the first bacterium to be isolated in pure culture and is further divided into *Lc. lactis* subsp. *lactis* and *Lc. lactis* subsp. *cremoris* (a sub-species is a closely related but distinct organism). *Lc. lactis* subsp. *lactis* grows in 4% NaCl, at 40°C and produces ammonia from the amino acid, arginine, whereas *Lc. lactis* subsp. *cremoris* does not. *Lc. lactis* subsp. *cremoris* is generally thought to give better flavored butter and cheese than *Lc. lactis* subsp. *lactis*, but definitive proof of this is lacking. As we shall see below, other species of LAB are also used in starters.

Up to the end of the 19th century, butter was the dominant dairy product and was made on farms from cream separated from milk by gravity. The cream was stored at room temperature for perhaps 3 or 4 days to accumulate sufficient quantities for churning. Storage allowed many LAB contaminating the cream or milk (and indeed other bacteria) to grow (there was very little refrigeration at that time) and produce the necessary lactic acid. In 1878, Gustaf de Laval invented the centrifugal separator, which allowed much faster separation of cream from milk and facilitated consolidation of butter production from farms into larger units, called creameries. As an alternative to waiting for the contaminating LAB

*From Farm to Table.* Alan Kelly, Patrick Fox, and Tim Cogan, Oxford University Press.
 DOI: 10.1093/9780197581025.003.0004

to grow, the farmer sometimes added a small amount of a previous batch of buttermilk to the cream (called back-slopping), which was then held at room temperature for several hours to allow the LAB to grow. The cream coagulated (due to lactic acid produced by this fermentation) and the butter produced was lactic butter, which had a quite different (stronger and more acidic) flavor to that of sweet cream butter, which is the more traditional butter that is produced in English-speaking countries and is made from non-fermented cream.

Around 1890, Wilhelm Storch in Denmark and Herbert Conn in the United States began to advocate the use of "natural" starter cultures to give the desired flavor and aroma to the butter. These bacteria were isolated from cream, and some of them were shown experimentally to produce clean, lactic flavored butter, but none of them produced the aroma of a good flavored butter for reasons that became clear in 1919 when a good starter culture was shown to be a mixture of two bacteria, *Lactococcus lactis* and *Leuconostoc* spp.

Commercial production of starter cultures in specialized laboratories began in Denmark in 1891, and by the end of that century were in use in almost all creameries in Denmark. The ultimate source of the LAB was most likely plant material fed to cows and/or contamination from the milking environment, particularly the milking equipment.

## Types of cultures

Starter cultures are divided into two main groups, mesophilic, with an optimum temperature for growth of ~30°C, and thermophilic, with an optimum temperature of ~42°C. Each of these, in turn, can be divided into mixed- and defined-strain cultures. Mixed cultures contain several strains of the same and, indeed, different species, while defined or single strains are individual known strains, which, confusingly, are generally used in mixtures of three or four strains.

Mesophilic mixed starters contain two bacteria, i.e., *Lactococcus lactis*, which metabolizes lactose to lactic acid and *Leuconostoc* species, which metabolize the small amounts of citrate present in milk and cream to diacetyl, acetate, and $CO_2$. Diacetyl is produced in small amounts, and both it and acetate are important components of the flavor of lactic butter, cottage cheese, and cultured buttermilk. $CO_2$ is responsible for the production of the holes or eyes that are characteristic of Edam and Gouda cheese. In 1936, it was shown that citrate could also be metabolized by some strains of lactococci. This organism was initially named *Sc. diacetilactis* but has since been renamed a citrate-utilizing (referred to as $Cit^+$ ) strain of *Lc. lactis*, because genetically it is very closely related to *Lc. lactis*. The ability to transport citrate into the cell is encoded by a piece of DNA not found on the chromosomes called a plasmid (see below). Leuconostocs grow poorly in milk but are able to grow well in association with lactococci, for reasons that are not clear.

Many traditional mixed cultures in use today are probably sub-cultures of sour milks that produced good quality butter and cheese at the end of the 19th century.

Mesophilic mixed cultures are ~90% strains of *Lc. lactis*, which are unable to utilize citrate ($Cit^-$ strains), and ~10% of strains that are able to utilize citrate ($Cit^+$) and are therefore considered to be mixed cultures.

Depending on the $Cit^+$ component, mesophilic cultures are divided into four groups:

- O cultures, containing no $Cit^+$ strains;
- D cultures, containing $Cit^+$ strains of *Lc. lactis*;
- L cultures, containing $Cit^+$ strains of *Leuconostoc* species; and
- DL cultures, containing $Cit^+$ strains of both *Lc. lactis* and *Leuconostoc* species.

Thermophilic mixed cultures almost always consist of two organisms: *Streptococcus thermophilus* with either *Lactobacillus helveticus* (for Swiss cheese production) or *Lb. delbrueckii* subsp. *lactis or Lb. delbrueckii* subsp. *bulgaricus* (for yoghurt production). Due to their distinctively different shapes, these are often referred to as the coccus and rod, respectively. New strains and species are being isolated regularly; e.g., *Sc. galolyticus* subsp. *macedonicus* was isolated from Greek cheese in 1998 and shows promise as a "new" thermophilic starter for dairy fermentations. Color-enhanced photomicrographs of some species of LAB used as starter cultures are shown in Figure 4.1, where the differences in shape are very evident. *Lb. casei, Pediococcus pentosaceous* and *Lb. brevis* are not used as starter cultures but are important members of the non-starter microflora of many cheeses (see Chapter 6) and are shown for comparison.

Bulk starters of the rod and coccus are generally grown separately for Swiss cheese manufacture but together for yoghurt production. The two organisms grow better together, because the proteolytic system of the lactobacilli produces amino acids from casein, particularly leucine, isoleucine, and valine, which stimulate the growth of the streptococcus, while the streptococcus produces $CO_2$ and formate (both probably from lactose), which stimulate the growth of the lactobacillus. This phenomenon, where one organism produces nutrients that stimulate the growth of another organism, is called symbiosis.

In Italy, France, and Switzerland, other types of mixed cultures, called natural whey cultures (NWCs), are used in the production of Parmigiano Reggiano, Grana Padano, Comté, and Swiss cheeses, all of which are normally made from raw milk. These cultures are derived mainly by back-slopping, as described earlier; in this case, some of the previous batch of whey is incubated

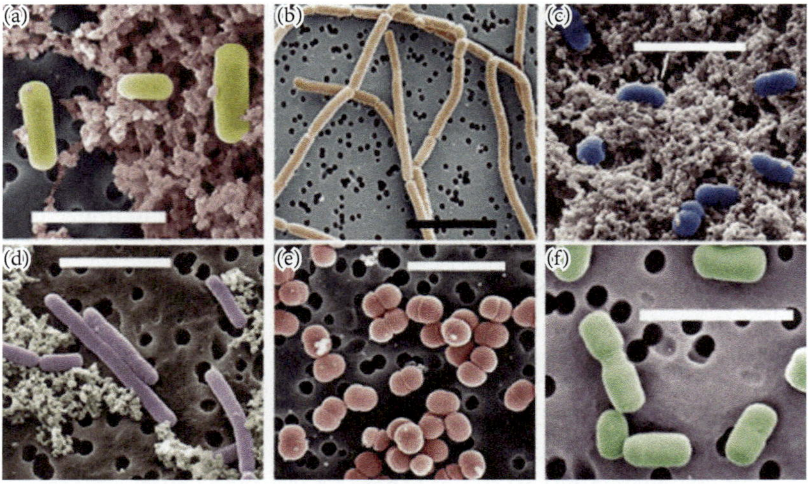


**Figure 4.1** Color-enhanced scanning electron micrographs of some starter and non-starter bacteria. (A) *Lactobacillus helveticus*; (B) *Lb. delbrueckii* subsp. *bulgaricus*; (C) *Lactococcus lactis*; (D) *Lb. casei*; (E) *Pediococcus pentosaceus*; (F) *Lb. brevis* (from Broadbent and Steel. 2005. *Cheese Flavour and Genomics of Lactic Acid Bacteria.* ASM News, 71, 121); Image credit to the American Society for Microbiology.

under prescribed conditions (e.g., 45°C for 18 hours) for use as the inoculum for the new batch of cheese. No special precautions are taken to prevent contamination from the cheese-making environment and, as a result, natural cultures are continuously evolving. The incubation conditions exert selective pressure on the types of bacteria that grow, and the composition of these cultures is very complex and variable. *Lb. helveticus* is the dominant organism in NWCs but one or more of the following organisms may also be present: *Lb. delbrueckii* subsp. *lactis*; *Lb. delbrueckii* subsp. *bulgaricus*; *Lb. plantarum*; *Lb. casei*; *Lb. paracasei*; *Lc. lactis*; *Sc. thermophilus*; *Ec. faecalis*; *Ec. faecium*; and *Leuconostoc* species.

A relatively new development in starter cultures is the use of defined strains. Their development was pioneered in New Zealand for Cheddar cheese in the 1970s. They are pure cultures, the characteristics of which are known, and are used especially in large Cheddar cheese-making plants. Mixtures of three or four strains are generally used, carefully selected on the basis of their ability to produce acid at 30°C and 37°C, lack of off-flavor production when grown in milk, and, most importantly, their ability to withstand attack by viruses that attack such bacteria, which are called bacteriophage (or phage for short). Each strain is checked daily for phage by growing it in milk in the absence or presence of filter-sterilized whey (as described below).

## Taxonomy of starter bacteria

Taxonomy is a scheme of classification in biology, based on the results of various tests, some of which were introduced in Chapter 3. In the case of starter LAB, their shape, Gram reaction, the number of different sugars fermented, the pathway of fermentation, whether they do not contain the enzyme catalase, and whether they grow at 10°C, 15°C, or 45°C are important tests used in classifying them. Starter LAB belong to several genera of bacteria including, *Lactobacillus*, *Lactococcus*, *Streptococcus*, *Enterococcus*, and *Leuconostoc*. These are all catalase negative, Gram-positive LAB. Lactobacilli are rod-shaped; all the others coccal-shaped. Most starter LAB, except leuconostocs (and some lactobacilli but not those generally found in starter cultures) ferment sugars by glycolysis in which the only product of the fermentation is lactic acid. For this reason, they are called homofermenters. The leuconostocs ferment sugars differently and produce other products, acetate and $CO_2$, in addition to lactic acid, from the sugar, and are called heterofermenters.

The taxonomy of lactococci, streptococci, and enterococci has gone through considerable change during the past century, due mainly to advances in the use of molecular techniques like nucleic acid hybridization and DNA sequencing.[1] In 1983, *Streptococcus lactis* and related streptococci were transferred to a new genus, *Lactococcus*, as *Lc. lactis* subsp. *lactis* and *Lc. lactis* subsp. *cremoris*;[2] and the so-called fecal streptococci were transferred to the genus *Enterococcus* as *Ec. faecalis* and *Ec. faecium*. In addition, the genus *Lactobacillus* has recently been divided into 24 new genera, including *Lactobacillus* itself, based on molecular, physiological, and ecological techniques.[3] These changes have affected changes in the names of some of the lactobacilli that are found in cheese and yoghurt. *Lb. helveticus* and *Lb. delbrueckii* subsp. *bulgaricus* have retained their old names, but *Lb. casei* and *Lb. paracasei* have been transferred to a new genus, *Lacticaseibacillus*, as *Lacticaseibacillus casei* and *Lacticaseibacillus paracasei*, respectively, and *Lb. plantarum* to a new genus as *Lactiplantibacillus plantarum*. The old names are retained in this chapter, as they are still frequently used in practice.

## Enterococci as starters

Enterococci are also LAB. They are commonly found in natural whey cultures and in freshly made and ripened cheese and are considered to have a positive effect on the development of cheese flavor. However, there is considerable debate as to whether they should be used in cheese-making since the major sources of many of them are animal and human feces.[4] Enterococci are thermophilic, since most strains grow at 45°C; they also grow at 10°C and hence can potentially grow during cheese ripening. They withstand pasteurization and some strains have been shown to produce bacteriocins (see below) during Cheddar cheese manufacture, which inhibit listeria. The common species are *Ec. faecalis* and *Ec. faecium*, but two new species were described recently, *Ec. italicus* and *Ec. lactis*, both isolated from Italian raw milk cheeses. These were susceptible to several clinical antibiotics and showed low virulence profiles, when screened for the presence of virulence genes, implying that *they* are a low health risk.[5] Unfortunately, there are no simple, biochemical, or physiological tests that will categorically separate lactococci from enterococci. Instead molecular techniques, like DNA sequencing, are needed to distinguish between them.

## Genes and plasmids

Like all bacteria, starter cells contain only one chromosome, made up of two complementary strands of deoxynucleic acid (DNA), each of which contains all the genetic information necessary for the metabolism, growth, and multiplication of the cell. This information is contained in genes, each of which encodes an individual protein. So, a gene can be defined as a section of DNA that contains the genetic code for the production of a protein.

Starter cells also contain pieces of DNA called plasmids, which are independent of the chromosome; the number of plasmids present varies between species, e.g., *Lc. lactis* strains generally contain between 4 and 7 plasmids, while *Sc. thermophilus* strains contain only 1 plasmid, if any. Many of the commercially important properties of starter cultures are encoded by plasmid genes, including the transport of citrate into the cell, phage resistance, and production of complex sugars called exopolysaccharides, and bacteriocins, and several of the proteins involved in the transport of lactose. Plasmids also encode the proteinase system that hydrolyses the milk proteins to peptides and amino acids, which are required for growth of the starter and are also precursors of many of the compounds involved in flavor development of cheese during ripening.

Plasmids are at least 1,000 times smaller than the chromosome and are easily lost on sub-culture; once this occurs, the properties encoded by that plasmid are also lost and the cells may grow poorly in milk as a result, e.g., if they lose the proteinase or lactose transport plasmids. For this reason, sub-culturing of starters should be limited to a few transfers. Plasmids are relatively easy to transfer between strains, and this is the basis of genetically manipulating commercial strains to improve their technological abilities, e.g., inserting phage-resistance plasmids into phage-sensitive starter strains.

## Differentiation of starter strains

Molecular analysis of differences in DNA composition is the main way to distinguish between strains of microorganisms. One of these techniques, called pulsed field gel electrophoresis (PFGE), has been used extensively on starter isolates. In this technique, DNA isolated from several colonies of the strains is hydrolyzed with an enzyme (called a rare-cutting restriction enzyme) that cuts the DNA into large fragments, which are then separated on a slab of gel material by an electric current (called electrophoresis) and visualized by staining the DNA fragments with ethidium bromide. In this technique, bands (representing fragments of DNA cut by the enzymes) that move to the same extent are considered to be identical, and isolates that differ by not more than 1 or 2 bands are considered to be virtually identical. These techniques can be used to study the starters being used for cheese-making and their relationships in exquisite detail. For example, in one study, the PFGE patterns of 289 strains of *Lc. lactis* subsp. *cremoris* from various sources divided the strains into 12 groups, indicating that many strains are related.[6] In thermophilic cultures, PFGE showed that all 9 cultures of *Sc. thermophilus* and 8 of 12 cultures of *Lb. helveticus*, from three US suppliers, contained only one strain and were therefore pure cultures; 4 other strains of *Lb. helveticus* were mixed cultures, containing either 2 or 4 strains.[7]

Sequencing of complete genomes of bacteria is now possible and is even more discriminatory than PFGE in determining differences between strains.

## Nutrition and metabolism of starters

All starter bacteria, and indeed LAB in general, are nutritionally fastidious organisms, requiring a fermentable sugar from which they derive energy (in the form of a molecule called adenosine triphosphate, or ATP, see Chapter 3) and several amino acids and vitamins for growth. Milk is therefore a very good medium for their growth, as it contains lactose as an energy source, protein, from which they can derive the amino acids essential for growth through the action of their proteolytic enzymes, and numerous vitamins.

Two different mechanisms are used to transport lactose. In lactococci, lactose is modified by addition of a phosphate group as it is transported into the cell, while in thermophilic cultures lactose is transported without such modification.

The metabolism of lactose and its constituent sugars to lactic acid also differs between starters. Lactococci convert 1 molecule of lactose to 4 molecules of lactic acid by glycolysis. In contrast, some strains of *Sc. thermophilus* and *Lb. delbrueckii* convert only the glucose portion of lactose to 2 molecules of lactic acid, because they are unable to metabolize galactose and excrete it in proportion to the amount of lactose taken up. This has implications for cheese-making, since starters should metabolize both sugars so that only low levels of sugar are present in the cheese. Leuconostocs and heterofermentative lactobacilli convert 1 molecule of lactose to 2 molecules of lactate, 2 molecules of ethanol, and 2 molecules of $CO_2$.

Lactic acid can exist as two different isomers, called the L and D forms, due to the presence of an asymmetric carbon atom (a carbon atom to which four different chemical groups are attached) in the molecule and are mirror images of each other. *Leuconotoc* species and *Lb. delbrueckii* produce D lactic acid and *Lc. lactis* produces the L form; some species, e.g., *Lb. helveticus*, produce both isomers.

The amino acids glutamate, methionine, valine, leucine, isoleucine, and histidine are required by most lactococci, and many strains of *Sc. thermophilus* and leuconostocs have additional requirements for the amino acids phenylalanine, tyrosine, lysine, and alanine All required amino acids are released from casein by the starter cell's proteolytic enzyme system.

Citrate in milk is metabolized to $CO_2$, acetate, and diacetyl by leuconostocs and $Cit^+$ strains of lactococci; $CO_2$ can also be produced from lactose by leuconostocs. The $CO_2$ from both sources is responsible for the small eyes found in Dutch type cheese, while diacetyl, which is produced only in small amounts, is responsible for the pleasant taste of cultured buttermilk, sour cream, and lactic butter.

Another important flavor compound, especially in yoghurt, is acetaldehyde. This is generally considered to be a product of carbohydrate metabolism, but it can also be produced from the amino acid, threonine, by the enzyme threonine aldolase.

Swiss-type cheeses are characterized by the presence of propionate and large holes, called eyes. Propionate is responsible for the nutty, sweet taste of these cheeses, and $CO_2$ is responsible for eye formation. These compounds are not produced by LAB but by *Propionibacterium freudenreichii*, selected strains of which are deliberately added to the milk with the starters and transform the lactate produced by the starter LAB during the early stage of cheese-making to propionate, acetate, and $CO_2$ during the initial stages of ripening:

$$3 \text{ lactate} \rightarrow 2 \text{ propionate} + 1 \text{ acetate} + 1\ CO_2$$

## Exopolysaccharides

Many strains of LAB produce exopolysaccharides (EPS), which are composed of complex polymers of sugar molecules. They are called homopolymers if only one sugar is involved (e.g., dextran, which is comprised only of glucose molecules) and heteropolymers if more than one sugar is involved. Through increasing viscosity, EPS can improve the mouth feel and creaminess of fermented milks. EPS-producing strains have also been used to improve the texture of reduced fat cheeses, which are often rubber-like in texture. A simple way to test whether a strain produces EPS is to determine if long strands of coagulated milk can be pulled from milk-grown cultures with an inoculating loop; individual colonies can be tested in a similar manner.

## Bacteriocins

Bacteriocins are small, heat-stable peptides, containing 20–60 amino acid residues, which are produced by many strains of bacteria and have the ability to inhibit the growth of other bacteria. Bacteriocins can have a broad spectrum of activity, inhibiting many different bacteria, or a narrow spectrum, inhibiting only closely related bacteria. Many LAB produce bacteriocins, including starter LAB (*Lc. lactis*, *Sc thermophilus*, *Lb. helveticus*, and *Lb. delbrueckii*), non-starter LAB (NSLAB) like *Lb. casei* and *Lb. plantarum*, and some *Bifidobacterium* spp. Some strains of lactococci produce several bacteriocins. The best known bacteriocin is nisin, which is produced by some strains of *Lc. lactis* and was discovered in 1928; *Lc. lactis* is now known to produce several other bacteriocins besides nisin.

Bacteriocins are divided into three major classes: Class I are also called lantibiotics (because they contain the uncommon amino acid, lanthioine); Class II have anti-listerial activity; and Class III are large heat-labile proteins. Bacteriocins act by producing transient pores in the membrane of the bacterial cell, which allow the leakage of various essential compounds from the cell into its environment, leading eventually to cell death.

One of the most commonly used bacteriocins in dairy applications is the Class I bacteriocin nisin, which is produced commercially as Nisaplin. Its use as an additive in various thermally processed foods is permitted in over 80 countries. In the cheese industry, nisin is used in processed cheese production to prevent the germination of spores of *Clostridium tyrobutyricum* and *Cl. butyricum*, which can convert lactic acid to butyric acid, carbon dioxide, and hydrogen gas ($H_2$). The butyric acid gives the cheese a rancid flavor and the $H_2$ causes considerable expansion of cheese volume and the development of large irregular shaped eyes. Nisin also inhibits several other groups of undesirable bacteria, including lactobacilli, staphylococci, micrococci, and listeria.

Nisin is a peptide that exists in five different forms, all of which show slight differences in amino acid sequence, and which may exist as dimers and tetramers (of two or four peptides acting together). The producing strains have an immunity system that prevents them from being inhibited by their own bacteriocin. Production of many bacteriocins is encoded by plasmids, which are relatively easy to transfer between different strains. This is particularly important for strains of *Lc. lactis* and other starters, since the bacteriocin-producing strain of *Lc. lactis* may not be a good acid producer, but transferring the bacteriocin gene to a good acid-producing strain overcomes this problem.

## Bacteriophage

Bacteriophage, or phage, are viruses that can multiply only inside a bacterial cell. They are much smaller than the host bacteria and can be "seen" only with an electron microscope. They generally consist of a head containing the DNA, a tail consisting of protein, and a base plate, generally containing spikes. They were first isolated for *Lc. lactis* in New Zealand in 1935 and have since been isolated for all starter bacteria. They are strain-specific, in that a phage for a strain of *Lc. lactis* will attack only that and closely related strains. Infection of a starter culture by phage can significantly upset cheese manufacturing schedules and, in extreme cases, result in complete failure of acid production or "dead vats." This effect is quite dramatic and is easily seen by comparing the growth of starter cultures in milk in the presence and absence of phage (Figure 4.2).

Phage multiplication occurs in either of two ways, the lytic and lysogenic cycles. In the former, the phage tail adsorbs onto special attachment sites, called phage receptors, on the surface of the host cell. A schematic of phage attacking

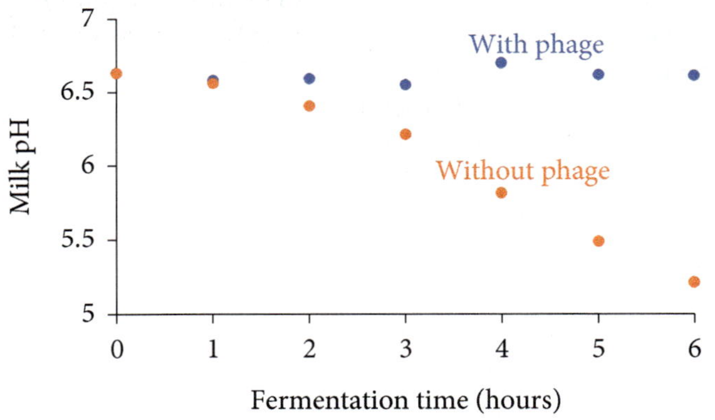


**Figure 4.2** Growth (as indicated by decrease in pH during fermentation) of a starter culture in the presence and absence of whey containing phage, which attacks that culture.

a bacterial cell is shown in Figure 4.3. This step requires the presence of calcium, which is present at high concentrations in milk. The phage then injects its DNA into the host, which immediately starts producing phage DNA and phage proteins rather than host DNA and host proteins. The phage DNA is packaged in the phage head and, when phage synthesis is completed, the cell lyses (bursts), releasing new phage particles, often as many as a 100 new phage particles per cell in about 1 hour, to restart the process by infecting 100 new cells. The newly infected cells will release another 10,000 phage (100 × 100) in a further hour, which will produce a million (10,000 × 100) phage in a further hour, and so forth. In this way, contamination with one phage can wreak havoc on a starter culture in a relatively short time, as phage multiplication is very rapid. Lysis is caused by a lytic enzyme, called lysin, which is encoded in the phage DNA, and which hydrolyses the cell wall of the host cell.

In the lysogenic cycle of phage multiplication, adsorption and injection of the phage are similar to those of the lytic cycle but, instead of phage multiplication, the phage DNA is inserted into the bacterial chromosome and multiplies with the host. This type of phage is also called a temperate phage, and under some, as yet ill-defined, circumstances can be induced to go through the lytic cycle to produce large numbers of phage. Lysogenic phage are thought to be the origin of phage in a cheese plant. However, many mixed cultures contain lysogenic phage

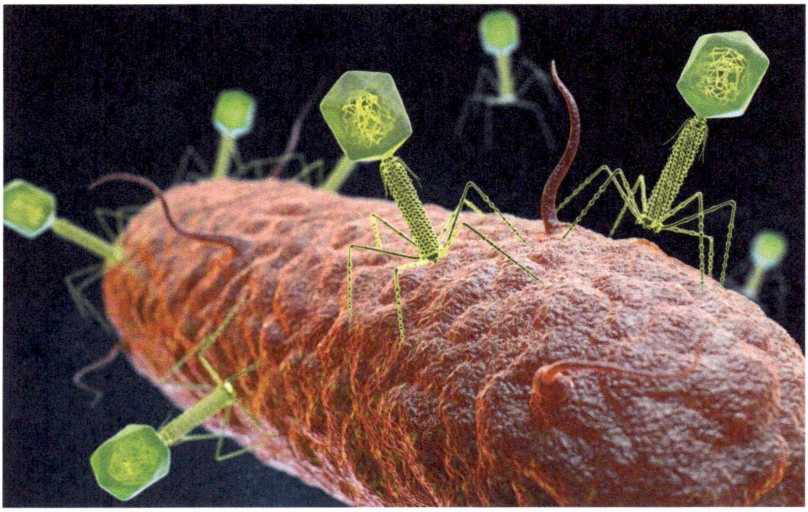

**Figure 4.3** Schematic showing the attachment of several bacteriophage to a rod-shaped bacterial cell. On attachment, the phage then inject their genetic material, which is encapsulated in their heads, into the bacterial cell (from iStock Images, with permission).

but no genetic relationship has been found between the DNA of the lysogenic phage and the DNA of the lytic phage that attack these cultures, implying that lysogenic phage may not be the source of lytic phage. Phage have also been found on walls, air, door handles, and tables, as well as in the cheese milk itself, indicating that they are ubiquitous in cheese plants.

A fully grown starter culture contains $10^9$ bacteria/ml and so a 100 L tank of starter contains about $10^{14}$ bacteria. One phage getting into such a tank could do untold damage. Therefore, the most important factor in producing good quality cheese is to ensure that the bulk starter is free of phage; this point cannot be over stressed, but it is often overlooked in commercial practice.

There are several important factors in controlling phage:

- use only a limited number of phage-unrelated strains of starter bacteria and use them in rotation;
- inoculate starters aseptically (i.e., under sterile conditions);
- use phage-inhibitory media (containing high levels of citrate and phosphate, which chelate the calcium necessary for phage to attach to the cell) for starter production;
- keep starter and cheese production areas physically separate from each other, with no direct access between them;
- add starter and rennet to the milk together, since the coagulum prevents movement of phage to non-infected cells and the more quickly it is formed the better;
- chlorinate cheese vats between fills, because chlorine is a very effective inactivator of phage.

Phage levels in starters and cheese whey should be determined daily in any well-run, quality-control scheme in modern cheese-making facilities.

## Other inhibitors of starter cultures

Mastitis is a common disease of the dairy cow udders and is routinely treated by infusing the infected quarter of the udder with an antibiotic, which invariably results in antibiotic residues in the milk and inhibition of the growth of the starter. Penicillin is commonly used, or one of its derivatives, and the milk from the infected animal should be withheld for 48 hours to reduce the risk of inhibition of the starter. Thermophilic cultures are much more sensitive to penicillin and cloxacillin (a derivative of penicillin) residues in milk and much more resistant to streptomycin than mesophilic cultures by almost an order of magnitude; both types of cultures have similar sensitivities to tetracycline. Today, this type of inhibition is less important than inhibition by phage.

## Propagation of starter cultures

Until perhaps 40 years ago, starter cultures were produced in cheese, butter, or fermented milk plants by progressive build up from small volumes of mother culture to larger volumes of bulk culture over several days. The mother cultures for this build up were obtained from laboratories that specialize in producing them in liquid or powdered form. The propagation medium was skim milk or 10% non-fat milk solids heated to 90–95°C for 30 minutes to inactivate any phage and bacteria that might be present, before cooling the milk to the incubation temperature. The reason for this time and temperature is that it suited the factory time sequence, since a mesophilic culture inoculated into the milk at, say 6 p.m., will be fully grown at 8–10 a.m. the following morning. It was then sampled, tested for activity and the presence of phage, and chilled to 4°C. Thermophilic cultures were propagated for 6–8 hours at 37–42°C before cooling to 4°C and testing for activity and phage. During the incubation, the cultures produce lactic acid, which results in a reduction in pH. An example of the growth of a mesophilic culture in 10% non-fat milk solids at 21°C is shown in Figure 4.4. The fully grown starter culture in milk contains about $10^9$ cells/ml. Chilled mesophilic cultures retained activity for 1–3 days and thermophilic cultures for at least a week after

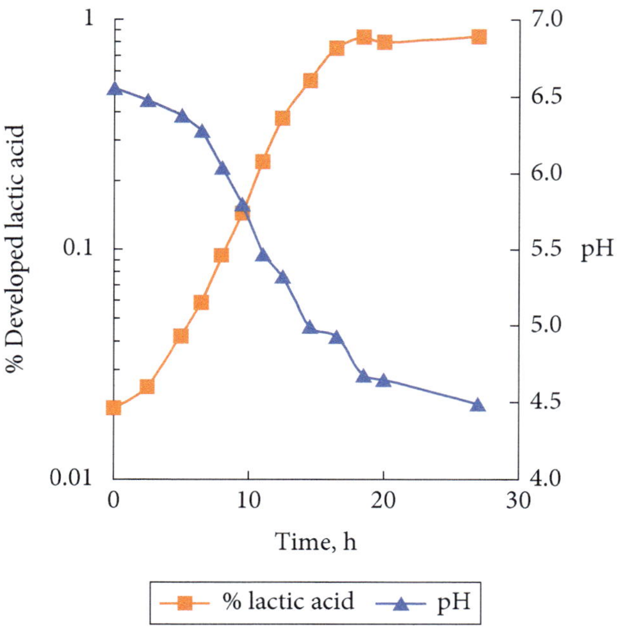


**Figure 4.4** Relationship between lactic acid production and pH in a mesophilic culture grown in 10% skim milk solids at 21°C (from Fox et al. 2017. *Fundamentals of Cheese Science*. 2nd edn. Springer, New York, with permission).

cooling. Sometimes in large cheese plants the medium for culture growth is a phage inhibitory medium containing milk solids, growth promoters, and extra citrate or phosphate to chelate the $Ca^{2+}$ necessary for phage attachment to the host cell.

Today, direct inoculation of milk in the vat with concentrated suspensions of frozen or freeze-dried starters, called direct-vat-set (DVS) cultures or direct-vat-inocula (DVI), is the norm. These come in volumes suitable for inoculation of 200–1,000 liters of milk and are particularly valuable for small cheese producers, because, although more expensive than traditional starters, they eliminate the need for sub-culturing and its attendant costs in terms of equipment, labor, and testing.

# 5
# Pasteurized and Long-Life Milk

## The diversity of dairy products

Having first considered the production and constituents of milk and then key aspects of the microbiology of milk, the following chapters of this book will focus on the production of the principal families of dairy products. Many of these evolved centuries or even thousands of years ago to solve the key challenges of preserving the nutritional value of milk in the form of safe and stable products, using very long-established methods of food preservation—fermentation, heating, drying, salting, and in some cases, smoking.

As a result, dairy products such as cheese, butter, and fermented milks have been produced in some form throughout human history, and the principles of their production in plants, large or small, today closely reflect long-standing principles, optimized and modified by decades of scientific research on the impact processing has on both the microbiological and chemical properties of milk.

A number of core processes are applied in different combinations in the modern production of dairy products—thermal processes such as pasteurization and ultra-high temperature (UHT) treatment, homogenization, separation (by centrifugation or membrane filtration), fermentation, evaporation, and spray-drying being the principal ones. The main processes applied to different dairy products are listed in Table 5.1; each product will be discussed in the following chapters, beginning here with fluid milk intended for consumption.

## The instability of milk

In many countries today, the main form in which milk is consumed is as liquid or market milk, and the basic processes for converting raw milk into these products (adjustment of fat content, pasteurization, and sometimes homogenization) are also common to the first stages of producing many other dairy products.

Liquid milk is a far more complex product category today than was once the case. In many countries (e.g., Ireland up to the 1970s), it was once common for milk—pasteurized but not homogenized—to be delivered in glass bottles with foil lids; this packaging is still found in some countries (e.g., in some parts of the United Kingdom) today.

*From Farm to Table*. Alan Kelly, Patrick Fox, and Tim Cogan, Oxford University Press.
 DOI: 10.1093/9780197581025.003.0005

**Table 5.1** The production of a wide range of dairy products from a small number of key dairy processes

| Process | Primary product | Secondary products |
|---|---|---|
| Centrifugal separation | Cream | Butter, butter oil, ghee, low-fat cream, cream cheese |
| | Skim milk | Skim milk powder, casein, protein ingredients |
| Heating | Pasteurized, UHT and sterilized milk | |
| Concentration (by heat or membrane) and drying | Evaporated milk, sweetened condensed milk | Milk powders, infant formulae |
| Enzymatic coagulation | Cheese (most varieties) | Processed cheese |
| | Rennet casein | Cheese analogues |
| | Whey | Whey powders, whey protein concentrates and isolates, lactose |
| Acid coagulation | Acid casein | |
| Fermentation | Cheese, yoghurt, quarg | |

In many countries today, however, complex multilayer packages called Tetra Paks have replaced glass bottles, and the contents are almost always homogenized. In addition, products with a range of fat contents, from skimmed milk (less than 0.1% fat) through low-fat (typically 1.5% fat) to full-fat (usually around 3.5% fat), are available. Many of these products are also fortified today with added protein, minerals, or vitamins, depending on prevailing local consumer demands and nutritional preferences. This product category (fluid dairy products) also includes cream products, for dessert or use in coffee, which have a fat content of 18–45%, depending on their particular application.

In addition to the traditional pasteurized products, which have a shelf-life of around 2 weeks in the fridge, in many countries the preference is for products processed using UHT treatment, which have a longer-life, a more "cooked" flavor, and a shelf-life of up to 1 year at ambient (non-refrigerated) temperatures (which can, at different times of the year, range from around or below 0°C in winter to maybe 40°C in summer).

Liquid milk is thus now a diverse and complex product category, but all these different products and formats evolved from a search for solutions to specific problems presented by raw milk, fresh from a cow or other dairy mammal.

## The instability of raw milk

Milk taken directly from a cow is highly unstable, for a number of reasons. The first visible change is fat separation, with a yellowish cream layer becoming visible at the top of the container in which the milk is stored, which will dissipate when milk is mixed or stirred, but returns relatively quickly (Figure 5.1).

**Figure 5.1** Raw milk, when left to stand for a few hours, separates to show a distinct cream layer (image thanks to David Waldron).

Less obviously, at least at first, microorganisms start to grow, as described in Chapter 3. Milk leaves the cow's udder at body temperature (around 37°C), the optimum temperature for the growth of many bacteria. While the interior of the udder of a healthy cow is sterile, unless the teats are infected by disease (e.g., mastitis), as soon as the milk emerges into the environment it becomes contaminated from a wide range of sources, including the cow's skin, the air, and particularly the milking machine and ancillary equipment. Hygiene on farms to reduce bacterial populations is obviously very important, but it is very difficult to prevent contamination with bacteria that can grow in the rich nutrient medium of milk. Cooling the milk to slow down bacterial growth is an urgent priority immediately after milking, but this changes the growth dynamics of bacteria in the milk in favor of psychrotrophic species such as *Pseudomonas* spp. These secrete enzymes that break down proteins and lipids and thus spoil the milk, causing changes in taste and in appearance (from destabilized proteins that may coagulate and separate from the milk).

In addition, pathogens may enter the milk (perhaps from an infected cow's udder or the environment), which may not cause visible spoilage but rather lead to significant health hazards for consumers. One hazard particularly associated with raw milk was tuberculosis (TB), caused by the bacterium *Mycobacterium tuberculosis*, which for many years was a major public health concern. In Ireland and elsewhere until the middle of the 20th century, TB was a major cause of death or long-term illness, with many people spending long periods in specialized hospitals, called sanitoria, to recover. The conditions used to pasteurize milk were specifically chosen to inactivate this bacterium.

For all these reasons, milk has long been processed before consumption as a liquid and today is consumed raw only under very limited circumstances (by the farm family or, in some cases, where allowed by law, within a certain distance of a farm but within a very short time of milking).

The first processing principle applied to raw milk thus relates to safety, as should be the primary consideration when processing any food before consumption. The second, once safety has been ensured, is to extend its shelf-life for the convenience of consumers (who then need to purchase milk less frequently) and greater distribution opportunities for the producer (who can sell milk over longer distances).

The other processing considerations in the treatment of milk concern its aesthetic properties, such as appearance. For example, to reduce fat separation (creaming), milk may be homogenized, while to adjust the level of fat present to meet the nutritional preferences of consumers, milk may be separated into skimmed milk and cream which are then remixed to achieve a specific final fat content, a process called standardization.

## Microbiological effects of heating milk

To understand the microbiological objectives of pasteurization, consider the objective, which is to produce a product that is safe to drink and has a reasonable shelf-life, perhaps 2–3 weeks, if kept refrigerated.

To attain these goals, it is important to first consider the populations of bacteria present in milk and the effect of heating on these (there could, of course, also be yeasts, molds, and viruses present). There are three key populations of concern here.

*Pathogenic bacteria*: These are the bacteria which cause illness or death to humans if consumed; their inactivation is obviously the key objective of any heat treatment. Species in this category come into the milk from the environment, milking machines, during transportation or handling, or from diseased cows; examples include enteropathogenic *Escherichia coli*, food poisoning *Staphylococcus aureus*, *Salmonella*, and *M. tuberculosis*. The means by which these bacteria are killed is heat, so a key question concerns their resistance to heat. For each species extensive knowledge of their characteristics in this regard has been developed through research.

Two key factors are involved, the temperature reached and the time for which this temperature is held. Generally, the effect of heat on bacteria gives rise to a graph which is linear (a straight line), if the logarithm of the number of survivors (where the very large numbers of bacteria are plotted as factors of 10, i.e., 1, 10, 100, 1000, etc.) is plotted against time. This is illustrated in Figure 5.2.

From such graphs, one can calculate the time necessary to kill 90% of the population, which will be the same regardless of the initial concentration of cells. This is termed the D-value and is usually stated in seconds or minutes. So, to kill 99.9% of a population of microorganisms requires a time equivalent to 3 D-values. A plot of the log of the D-values against temperature is also linear and the slope of this curve is called the z-value (measured in minutes or seconds), which is the time necessary to cause a 10-fold difference in the D-value.

For example, 15 seconds at 65°C might be sufficient to kill 90% of the *E. coli* cells in milk. This means that, if milk is heated for this time at that temperature, any cell will have a 1 in 10 chance of surviving; if the milk is heated for 30 seconds (i.e., 2 × 15 secs) or 60 seconds (i.e., 4 × 15 sec), the chance of a cell surviving decreases to 1 in 100 or 1 in 10,000, respectively. By knowing the likely numbers of bacteria that might reasonably be found in milk, it is possible to determine the heating conditions required to give essentially no chance of any cells surviving and causing a health risk.

Such an evaluation could be conducted for all pathogens that might be present in milk and these bacteria ranked in order of their resistance to heat. In the case of milk, the most resistant pathogenic bacterium was long thought

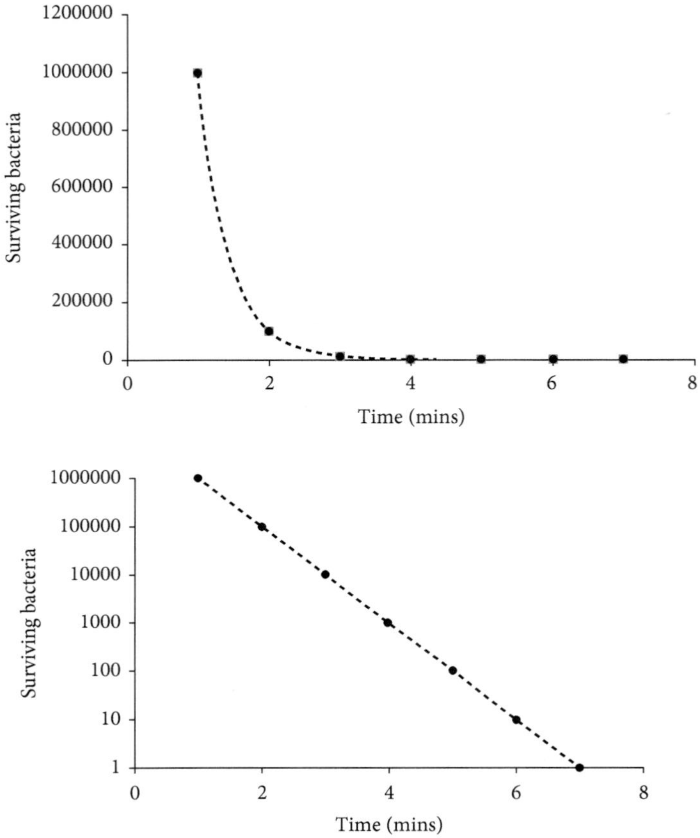


**Figure 5.2** The death of bacteria as a function of time during heating at a constant temperature, plotted as (top) the arithmetic number of bacteria surviving and (bottom) the logarithm of the number of bacteria surviving. The time to reduce the number of bacteria by a factor of 10 can be obtained readily from the bottom plot and is called the D-value at that particular temperature.

to be *M. tuberculosis*, with the conditions required to give a high degree of inactivation of this bacterium (5 D-values, meaning any cell has a 0.0001% chance of survival) being determined to be 15 seconds at 72°C. These then became the conditions used for typical commercial pasteurization (known as high-temperature short-time, HTST, pasteurization) of milk. If *M. tuberculosis* was the most heat-resistant pathogen present, it follows that if it is inactivated, then all bacteria of lesser resistance to heat should also be killed. These conditions remained in common use until relatively recently, as will be discussed later.

*Spoilage bacteria*: This group includes bacteria that do not present a health hazard, but rather spoil the milk as they grow, making it less acceptable to consumers. This might be due to the production of acid by lactic acid bacteria, or the production of enzymes that break down fats and proteins, with their impact on proteins causing the milk to coagulate. Of particular relevance here are *Pseudomonas* spp., which are psychrotrophic bacteria (meaning they can grow at a low temperature), some strains of which produce such enzymes. *Pseudomonas* spp. are relatively susceptible to heat and are easily killed by pasteurization, but, if they had a chance to grow and multiply before the heat treatment, some of the enzymes they produce are very heat-stable and cause problems long after the bacteria that produced them have been inactivated.

If *Pseudomonas* are found in pasteurized milk, this indicates that either the correct heat treatment was not applied or that cells contaminated the milk after heating, so-called post-pasteurization contamination (PPC). Much of the improvement in the shelf-life and quality of pasteurized milk in recent decades has actually come from reduction in PPC, particularly due to the shift from using glass bottles to Tetra Paks, which are much easier to fill and seal without the risk of such contamination.

For other spoilage bacteria, their heat resistance and likelihood of surviving pasteurization can be evaluated, as discussed above for pathogens, and it will be seen that pasteurization extends the shelf-life of milk through a significant reduction in their numbers. It then simply takes far longer for them to grow to levels where they impair the quality of the milk. To cause visible change in milk generally requires numbers in excess of $10^6$ (1 million) bacteria/ml of milk.

*Spore-forming bacteria*: As discussed in Chapter 3, the hardest bacteria to kill by heat are those that form spores, which are very resistant to heat. This is where the key role of refrigeration of pasteurized milk comes into play, as these species usually do not grow at refrigeration temperatures and so remain dormant, being ultimately killed by our digestive juices when milk is consumed. So, a *Clostridium*, if present as spores, will remain in the milk throughout refrigerated storage, but will not present the threat it would if it were growing. The most problematic bacterium of this type is *Bacillus cereus*, which is both a spore-former and a psychrotroph, so survives pasteurization and can germinate and grow in milk at temperatures around 4°C, secreting enzymes that cause the kinds of undesirable changes consumers regard as spoilage. In modern production of fluid milk, spoilage by *B. cereus* is the most likely cause of deterioration, but this happens sufficiently slowly at refrigeration temperatures to give a reasonable shelf-life of perhaps 2–3 weeks, if the starting raw milk is of good quality and so has few of these bacteria present.

A notable point coming from this discussion is that pasteurized milk is actually preserved and made safe for consumption by a combination of three factors:

1. heat treatment to kill pathogens and reduce the number of spoilage bacteria;
2. packaging, which prevents recontamination after that heat treatment and during storage; and
3. refrigeration to slow down growth of surviving spoilage bacteria and prevent spores from germinating.

While pasteurization is often defined only in terms of the first of these, in reality it is critical that all three apply, and so pasteurization is an example of what is called a *hurdle process*, where multiple factors or operations work together to make a food product safe.

## Emerging threats and recent developments in pasteurization

As mentioned above, the classical "target" organism for pasteurization was *M. tuberculosis*, and the introduction of widespread pasteurization of milk was highly successful in controlling the spread of TB in many countries.

However, the efficacy of pasteurization in protecting public health has come under new scrutiny in recent years, due to the emergence of a new threat from a bacterium related to *M. tuberculosis*, *Mycobacterium avium* subsp. *paratuberculosis* (or MAP). This bacterium causes a disease in cattle called Johne's disease, characterized by intestinal discomfort and weight loss. Crohn's disease in humans is broadly similar and some initial studies suggested that individuals who suffered from Crohn's disease were marginally more likely to have MAP in their intestines than those who did not. This led to the question of whether cows with Johne's disease could shed the MAP bacteria in their milk and transmit it to humans, who might contract Crohn's disease as a result; it is now generally accepted that there is no link between the bacteria and Crohn's disease.[1]

The immediate question then was whether pasteurization inactivated MAP as, if it did, this would mean that the presence of the bacterium in raw milk was irrelevant, as the heat treatment would kill it, and hence it could not cause problems for humans.

Answering this apparently simple question became a prolonged and somewhat controversial topic in dairy science. To gather data on the heat inactivation of MAP, the classical approach would be to find milk in which it was present, or more likely add it to milk at known levels, heat that milk for different times and at different temperatures, and count the surviving bacteria. For a typical bacterium, the incubation period for agar plates is 48 hours, and so a relatively quick answer can be found. However, MAP grow very slowly on plates, and obtaining a count can take up to 6 months, which makes studies on MAP and their survival very tedious and slow.

Initial studies suggested that traditional pasteurization at 72°C for 15 seconds was not quite sufficient to kill MAP, because the cells of that bacterium have a waxy or "sticky" cell wall and so tend to form clumps when they grow. When a clump of cells is heated, the outer bacteria might be killed but the inner cells of the clump are insulated, and so survive and grow.

On the basis of these studies, many processors changed the severity of their treatment, increasing the temperature of heating to 74°C, or the time for which milk is held at the maximum temperature to 30 seconds, or both.

However, the initial experiments on heat inactivation of MAP were criticized because they involved heating milk under relatively gentle conditions whereas, in industrial pasteurization, the highly vigorous mechanical treatments involved in pumping milk through the pasteurizer, plus processes like homogenization in particular, would break up the clumps of cells, making all of them susceptible to heat and less likely to survive.

Thus, years after the first reports and arguments emerged, the question of inactivating MAP by pasteurization remained somewhat contentious. However, there is no doubt that many processors have increased the severity of their heating processes just in case, as this is clearly to be preferred when there is even a low risk of any possible threat to human health occurring.

One important lesson learned from the episode, however, is that science can never assume it knows everything about any topic, as there is always a chance that new information can emerge and overturn what was assumed to be (up to that point) complete understanding of the matter.

## The physical properties of fat in milk

As discussed in Chapter 2, the fat in milk occurs in the form of an emulsion of fat globules that are less dense than the surrounding fluid and thus, under the influence of gravity, rise to the surface of the milk to form a cream layer. The rate at which they separate can be predicted mathematically from Stoke's law, which states that the rate of movement of the globules is dependent on a few simple parameters (as discussed in Chapter 2).

The first parameter is the size of the fat globules: the larger the globules, the faster they rise. This is actually the most powerful factor, as the rate at which separation takes place is proportional to the square of the radius of the fat globules. If one fat globule is twice as big as another fat globule, it will separate at four times the speed.

The second parameter is the difference in density between the fat globules and the surrounding fluid, which is what causes the globules to rise, as fat, being less dense than the surrounding fluid and immiscible with water, rises to the surface.

The third parameter is the force that is causing the separation to occur, in other words gravity. Really, rather than the fat globules rising, what is actually happening during separation is that the denser substance, water, is being pulled downward more rapidly than the fat, and so it sinks while the cream appears to rise. This is what happens if fresh raw milk is allowed to stand after milking, over the course of a couple of hours. However, the process can be greatly accelerated by enhancing the effect of gravity with the application of centrifugal force, in essence spinning the milk rapidly. The faster the speed at which it is being spun, the greater the effective force of gravity, and the more rapid is the separation of cream from milk.

The final factor that affects the rate of separation of oil from water is viscosity. However, in this case, the relationship is negative, and the more viscous the fluid surrounding the fat globules the more slowly they will rise.

These four factors (globule size, gravity or its more powerful counterpart, centrifugal force, the density difference between fat globules and the surrounding liquid, and the viscosity of that liquid) together underpin the key principles of separation, standardization, and homogenization of milk, three key steps in the production of fluid consumer milk.

For example, the most powerful way to reduce the rate at which fat globules separate, and prevent the milk from creaming or slow the rate at which it happens, is to reduce the size of the globules, which is what happens when milk is homogenized. On the other hand, the easiest way to speed up the separation of the cream from milk is to centrifuge the milk, thereby enhancing the impact of gravity.

Just as Stoke's law predicts the rate at which a fat globule rises to join a cream layer, it equally predicts the rate at which a heavy solid particle falls through a liquid. In this case, the solid particle is *more* dense than the surrounding liquid and instead of rising sinks. The rate at which it does so is entirely predictable (depending on size, density difference, and fluid viscosity) and can again be accelerated by centrifugation, as can be done in types of milk separators called clarifiers, which may be used to remove cells, dust, and other solid contaminants from milk.

## Homogenization

As discussed above, a key way to reduce the rate of separation of fat is reducing the size of the milk fat globules. This is achieved through the process of homogenization, which involves mechanical disruption of the milk fat globules by forcing them under pressure through a narrow opening where the forces experienced are so significant that the fat globules are torn apart into large numbers of much smaller globules.

A homogenizer contains a narrow valve through which a pump forces milk under pressure (see the schematic diagram in Figure 5.3). There is a small opening in this valve, and in being forced through it every globule is subjected to very severe forces, pulling, deforming, and tearing them as they are squeezed through a gap far smaller than they are. The modern milk homogenizer was developed by a Frenchman, August Gaulin, in 1899, who displayed his invention at the 1900 World Fair in Paris.

For homogenization to work efficiently, the milk fat globules must be easy to disrupt. At refrigeration temperatures, milk fat is largely solid, and it is very difficult to subdivide a solid material into smaller portions—think of ice, for example, compared to water. So the milk or cream must be warmed to a temperature at which the fat is liquid, i.e., above 40°C, where they will rupture more easily when forced through a very narrow gap.

The integrity of fat globules is maintained by the milk fat globule membrane (MFGM). During homogenization the mechanical forces applied to the milk rupture this membrane and split each fat globule into multiple smaller spherical globules, many of which are not covered by a protective membrane. A suitable visual analogy here is something the size of a soccer ball being divided

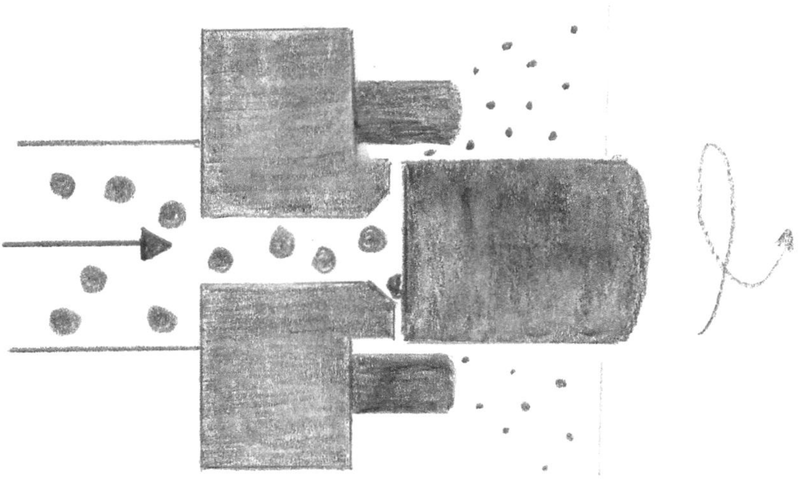

**Figure 5.3** Principle of operation of a homogenizer, showing the milk fat globules entering at the left of the diagram and being disrupted by their flow through the narrow gap surrounding the valve (image thanks to Anne Cahalane).

into a large number of smaller balls, maybe the size of table tennis (ping pong) balls. Taking the diameter of a football as 22 cm and that of a table tennis ball as 4 cm, and using the equation for the volume of a sphere, it can be calculated that each football will yield 166 table tennis balls. Using instead the much smaller sizes of fat globules in raw and homogenized milk (4 and 0.2 micrometers in diameter, respectively—1 micrometer is one thousandth of a millimeter) it can similarly be shown that each fat globule gives rise to 8,000 smaller globules.

The creation of so many new globules also greatly increases the surface area of fat that needs to be stabilized in the aqueous environment of milk (by a factor of 20 or so), and there is insufficient natural emulsifier (the MFGM) present to cover the newly formed globules. For a component to act successfully as an emulsifier, it needs to be amphiphilic, with both hydrophobic (water hating) and hydrophilic (water attracting) properties. Milk proteins (especially the caseins) have this property and, following homogenization, a small amount of the casein in milk associates with the fat globules to act as an emulsifier. To picture how this happens, note that the part of the homogenizer within which this happens is an extremely turbulent and violent place, with globules and casein micelles moving at high speeds due to the fluid velocities, mixing and mechanical forces (literally involving collisions with other globules or micelles and the metal walls and components of the homogenizer). When micelles and globules meet, the micelles fragment and become attached to the globule surface.

This process works remarkably well during homogenization, and results in the globules becoming coated in a mixture of original membrane material (some) and casein (mostly). However, the process isn't totally efficient, and the globules that form still have a tendency to clump together, so the homogenized milk is typically passed immediately through a second restriction valve (but less constricted than the first stage and operating at a lower pressure) just to break up these clumps or clusters (clusters have a larger size and so, according to Stoke's law, rise faster than individual globules).

The product then contains fat globules that are tiny relative to the original globules in raw milk, and so separate as cream far more slowly. In addition, due to their new protein coating, these globules are more dense, and so there is less of a difference between their density and that of the surrounding fluid, further helping to slow their rate of separation.

A decrease in diameter by a factor of 20 reduces the rate of separation by a factor of 400 (since the rate is proportional to the square of the size), according to Stoke's law. So, if milk separates into skim milk and cream in 12 hours under gravity, the same milk will take 200 days to separate if homogenized, and in practice much longer due to the increase in globule density. This means that homogenized UHT-treated milk, for example, can remain stable for months without showing a cream layer.

## Mechanical separation of milk

While leaving milk to stand will result in cream separating into a layer within a few hours, Stoke's law predicts that this can be accelerated by using centrifugal force. This was first commercially applied around the mid-19th century, and the first successful continuous machine for separating milk was developed by Gustaf de Laval in 1878. In 1883 de Laval and a business partner formed a company called AB Separator, which became Alfa-Laval and, later, one of the world's largest manufacturers of dairy processing equipment.

A modern milk separator is very efficient at separating phases of milk with different densities, whether this is removal of solids (e.g., animal cells, dust) from liquid milk or separating milk into cream and skim milk, by using centrifugal force. A schematic of a modern milk separator is shown in Figure 5.4. The material to be separated is fed from below (at A in the diagram) into narrow spaces between conical disks arranged in a stack (B), separated from each other by small metal stubs. The speed and dimensions are optimized such

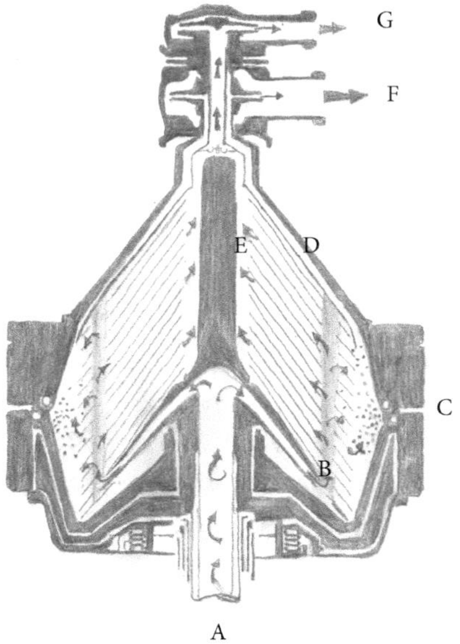


**Figure 5.4** Principle of operation of a separator. Whole milk enters a stack of rapidly rotating disks from below (A) and flows through holes along the periphery of each disk (at B). Due to differences in density, solid particulates are thrown to the outermost point of the chamber (C), skim milk flows along the outside of the disks (D), and cream flows inward (E). The separated skim milk and cream leave the separator separately (at G and F, respectively) (image thanks to Anne Cahalane).

that heavier materials are thrown against the outer walls, while the less dense fluid is directed toward the center of the stack and into channels from which, now depleted of the trapped solids, it exits the chamber. The solids that remain behind are eventually removed once they accumulate to a high level (at C).

If two liquids of different densities (skimmed milk and cream) are being separated, the whole milk enters the disk stack around half way along the surface of each disc, and the less dense material (the cream) is pulled inward toward the center (E), along the upper face of the lower disk; the heavier skimmed milk flows outward toward the outer walls of the chamber (D), along the lower face of the upper disk, and both leave the separator by different channels. So, one fluid flows in (whole milk) and two flow out (skimmed milk and cream, at points G and F, respectively).

The fat content of the cream exiting the separator can be easily checked (by measuring the density of the cream, which is directly related to the level of fat present) and a certain amount of the cream added back into the skimmed milk stream to achieve the desired fat content for the product or application in question (by *standardization*). For example, milk with 3.8% fat might be separated into cream (40% fat) and skimmed milk (less than 0.1% fat, in the form of very small fat globules that are too small to separate under the forces applied, as Stoke's law predicts), and then some of the cream added back into the skimmed milk to give a final level of 1.5% fat for low-fat milk for sale to consumers. The remaining cream may be packaged and sold for dessert applications, or else used for making butter.

## The pasteurization of milk

When the fat content of milk has been adjusted, by standardization, and the product has probably been homogenized, the key part of the process that remains is the actual heat treatment. The order in which these steps is conducted is important, with separation and then standardization followed by homogenization and finally pasteurization. The reason for the first two steps being in that order is that, according to Stoke's law, homogenized milk is exceptionally difficult to separate even using centrifugation, due to the very small size of the fat globules. Heat treatment comes last very much because of food safety.

A key approach in controlling the safety of most food processes today is called HACCP (Hazard Analysis of Critical Control Points), a system developed by NASA (the US National Aeronautics and Space Administration) in the 1960s to ensure the safety of food for astronauts on space missions. This involves identifying the critical control points (CCPs) of a food process, which are the steps where the greatest opportunity for contamination to occur exists or where there is the greatest risk of a problem if the step is not applied correctly. In the case of pasteurizing milk, the CCP is the heat treatment step, as this is the only place where bacteria and other microorganisms are eliminated. HACCP practice then states that, following such a CCP, as little as possible should happen that is likely to reintroduce risk or hazards. Centrifugation and homogenization, if applied after (downstream of) the heat treatment step could potentially represent risk points at which recontamination could occur, and would then need to be operated in a sterile or aseptic manner, which is more technologically challenging. Placing them before the heat treatment step, on the other hand, means that any microbial cells entering the milk at that point will be rapidly eliminated by heat and so do not present a threat. This consideration then determines the order of processing steps.

Other possible examples of CCPs might include recording the temperature reached by raw milk after a specific length of time, e.g., 4 hours during cooling on the farm, or measuring the pH after a prescribed length of time (e.g., 5 hours) in the manufacture of cheese.

In some cases (such as UHT treatment, as will be discussed later), a product that is more stable over the long its life (up to a year) is obtained by applying homogenization after the final heat treatment. In that case, special homogenizers (operating under aseptic, or sterile, conditions) must be used, to avoid recontamination.

The first process used to actually pasteurize milk was introduced in Denmark in the 1870s. It became compulsory there in 1890 to control the spread of TB among cattle through the feeding of raw milk. Shortly thereafter (1893), depots for the distribution of pasteurized milk for feeding infants were established in New York City. Early pasteurization processes typically heated milk to around 80°C for a very brief period (flash pasteurization). In some cases, especially in small processing plants, a process called low-temperature/long time (LTLT) heating may still be used, where milk is heated to 63–65°C for 30 minutes in a vat with a heating jacket.

In large-scale modern plants, however, the process applied is HTST pasteurization, at 72–74°C for 15–30 seconds, first developed in England by the APV Company in 1923. In terms of the efficiency of inactivating the bacteria of concern, this is exactly equivalent to the LTLT process, as increasing temperature greatly increases the efficiency of killing bacteria, and so requires a shorter time to have an equivalent effect.

The types of equipment used industrially for heating liquids such as milk are called heat exchangers, because they involve hot and cold liquids exchanging heat when brought into close proximity, as heat flows from the hot to the cold liquid according to the laws of that branch of physics called thermodynamics, which state that nature's response to a difference in temperature is to work to eliminate that difference.

In the case of pasteurizing milk, the system used is called a plate heat exchanger, the core element of which is a compact set of rectangular metal plates clamped into a frame and held apart by rubber gaskets, with fluids flowing along the gaps between the plates (a sample plate from a plate heat exchanger is shown in Figure 5.5). The plates are essentially arranged in pairs, with hot and cold liquids flowing between alternating pairs of plates, as shown in Figure 5.6. There are openings at each of the four corners of the plates which line up through the stack of plates to form tubes through which fluids can be directed, with the rubber gaskets around the holes on each individual plate forming barriers that can either direct a fluid to flow across a certain plate and leave through an "open" hole at the other end, or not. The surface of each plate is corrugated, which increases both the surface area across which heat can be exchanged and the turbulence of the fluid flowing across it, increasing the mixing and efficiency of transfer of heat.

A schematic diagram of the operation of a plate heat exchanger pasteurizer is shown in Figure 5.7. When raw milk enters the heat exchanger, it is heated through a process called regeneration by outgoing pasteurized milk; this is very efficient as it significantly reduces the amount of heating and cooling required. By absorbing heat from pasteurized milk, the raw milk is heated to 65–68°C (the process of regeneration is not 100% efficient, and some energy is always lost in the system).

In most operations, the process of regeneration is usually split into two parts and when the milk reaches 40–50°C, it is directed out of the pasteurizer, as this is the temperature at which separation, standardization, and homogenization are applied. These processes work best when the milk fat is liquid—as it is in this temperature range—and so their integration into the pasteurization process saves a separate set of heating and cooling operations. While the first part of the regenerative heating process is thus applied to whole raw milk, the second part heats standardized homogenized milk.

As the process of regeneration is not 100% efficient, a final heating step to bring the milk temperature to the pasteurization temperature is required, using hot water flowing between the plates adjoining those through which the milk is flowing. The milk then enters a holding section, typically an external coiled tube of dimensions that ensure the milk takes the required time (in the range 15–30 seconds) to pass through, based on its flow rate.

The milk then enters a critical part of the heat exchanger, the flow diversion valve, where an electrical thermometer checks its temperature against the preset target pasteurization temperature. The milk, following this test, can take either of two paths. If the temperature is too low, it will be rejected and flow along a tube

**Figure 5.5** A pasteurizer plate, showing the distinctive corrugations and positioning of the black gaskets, which block the flow of liquids on the right-hand side of the plate, but open channels on the left-hand side allow liquid to enter and flow along the plate.

that returns it to the start of the process, as it has not reached the conditions to ensure safety; machine operators will be informed and corrective actions taken to determine the cause of the problem. If, on the other hand, the temperature is found to meet the target, the milk is returned to the regeneration section of the pasteurizer, where it surrenders its heat to the incoming raw milk. By law, the

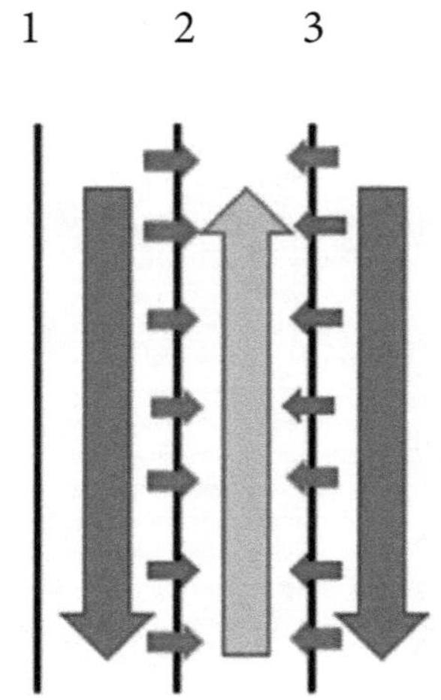


**Figure 5.6** Simplified flow pattern through a pasteurizer, with fluids flowing through alternating gaps between four plates (the black lines). The cool fluid (blue arrow) flowing through the middle pair of plates (between plates 2 and 3) is heated by heat flowing across plates 2 and 3, with which it is in contact from the hot fluid flowing in the gaps between plates 1 and 2 and plates 3 and 4 (red arrows). The liquids are flowing in opposite directions, which is more efficient in optimizing the transfer of heat and is called counter-current flow.

temperature in the flow diversion valve must be recorded to allow later validation in case of suspected safety issues with milk produced on a certain day.

One potential safety risk in the regeneration process is the possibility that raw milk might contaminate the pasteurized milk, for example through a tiny hole or crack in a plate on either side of which the two fluids are flowing. This risk is minimized by using a pump (called a booster pump) to increase the pressure on the pasteurized milk slightly, which means that, in the event of a leak, pasteurized milk would flow into raw milk, which is not a health risk, unlike raw milk flowing into pasteurized milk.

After this pump, the milk is cooled by regeneration, which again (because it is not 100% efficient) does not cool the milk down to the necessary 5°C, but perhaps only to 10–12°C, and so cold water is used in the final section, again flowing along the alternate pairs of plates, to bring the temperature down to 5°C, but, like the hot water used for final heating, this is only a small fraction of that which would be required were regeneration not used. The milk then exits the pasteurizer for immediate packaging or held in an intermediate tank until this is done.

In this way, a plate heat exchanger pasteurizer is a very efficient, safe (thanks to the use of the flow diversion valve and booster pump), and high throughput

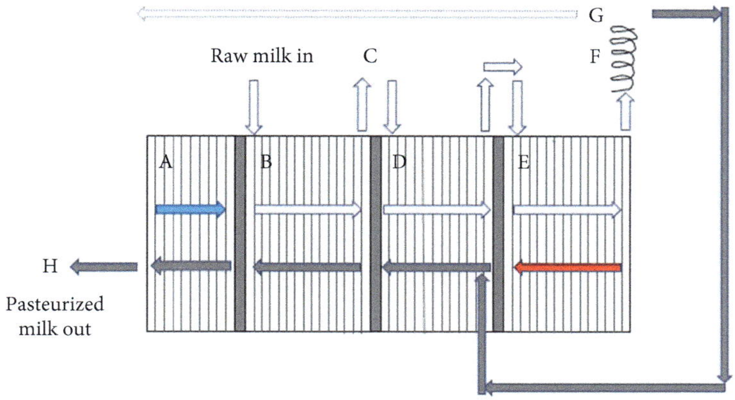


**Figure 5.7** Flow of milk through a plate heat-exchanger pasteurizer. Raw milk (white arrows) enters the regenerative heating sections (B and D), to be heated by outgoing pasteurized milk (gray arrows). Milk may leave at point C to be separated and homogenized, because it is at the optimum temperature for these operations during the regenerative heating process, before re-entering the pasteurizer to continue the heating process. Final heating takes place with hot water (red arrow) in section E, before the milk enters a holding tube or coil (F). Immediately after this, the temperature is checked and, if satisfactory, the milk passes into the regenerative sections (B and D) to lose heat to the incoming raw milk, followed by final cooling against chilled (blue arrow) water in section A and leaving the pasteurizer for further processing or packaging at H. Milk that has not reached the target pasteurization temperature at point G is rejected and returned to the start of the process while the issue causing this problem is addressed.

system, through which very large volumes of milk can be safely processed every day. A picture of a small pasteurizer is shown in Figure 5.8.

## Evaluation of the efficiency of pasteurization

How does a processor know if their process is operating effectively and achieving the required goals of pasteurization?

Processors must monitor their processes very closely, checking that the required temperature has been achieved, and indeed are required by law in many countries to maintain records to show that this is the case - another critical control point. The time for which the milk is held at the target pasteurization temperature is also checked but is determined by the dimensions of the holding section

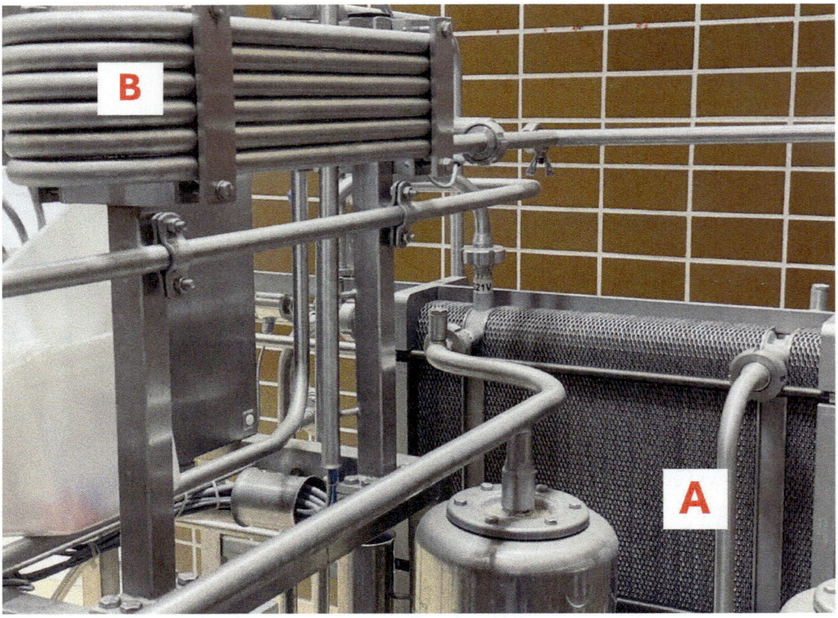


**Figure 5.8** A small pasteurizer, showing at right the plate heat exchanger with a large number of plates clamped into a frame (A), and at upper left the coiled holding tube (B).

of the pasteurizer and the flow rate of the milk. However, additional evidence is required to prove that milk that has been through the process is safe to consume.

If the target of the process was to kill a particular bacterium like *M. tuberculosis*, on the basis that this is the most heat-resistant pathogenic bacterium present, then it might make sense to test the milk after treatment to verify that it did not survive. However, such an approach has many drawbacks, including the complexity and slow speed of microbiological testing, as well as the fact that counting *M. tuberculosis* is difficult. A greater limitation, however, is how to interpret a negative result, as this could actually mean two quite different things: (a) that the bacterium was present and killed by the process, or (b) that the bacterium was never present in the first place, if good quality raw milk from healthy cows was being processed. If the second scenario was the case, then the raw milk would have tested negative for the pathogen, but this gives no information whatsoever as to whether any other bacteria of concern could still be present.

For these reasons, rather than counting bacteria, an ideal indicator of process efficiency would be:

(a) easy to measure;

(b) safe to measure in a food environment;
(c) always present or detectable in raw milk; and
(d) never present or detectable in pasteurized milk.

Luckily, milk contains exactly such an indicator in the form of the enzyme, alkaline phosphatase. This enzyme has no known function in milk that would impact the quality of milk or dairy products, compared to, for example, enzymes that break down lipids or proteins and so change the flavor or structure of dairy products.

Alkaline phosphatase is nonetheless of huge significance in dairy processing, not due to its presence but due to it having, coincidentally, a similar resistance to heat as *M. tuberculosis*. In raw milk, the activity of alkaline phosphatase can always be measured using a simple test in which milk is mixed with a chemical (called a substrate) that, when acted on by the enzyme, breaks down to a colored or fluorescence-emitting product, which is then measured using a suitable instrument in a matter of minutes or even seconds. If the milk has been pasteurized, however, the enzyme is inactivated, and no fluorescence is emitted when milk and substrate are mixed. Today, this test is used routinely in processing plants to validate the efficacy of pasteurization and to determine if milk has been pasteurized.

This principle of using an alternative (so-called surrogate) measurement which is easy to perform but yields critical information about something else that is much harder to measure is very common in food analysis. Many of the surrogates used are enzyme activities, as these are usually easy and fast to measure. It is simply a question of finding one in the food under consideration that is inactivated by the heat treatment for which an indicator is required. Many different enzymes in milk have been intensively studied for this reason.

When cream is bring pasteurized, it is usually heated to a higher temperature than milk of regular fat content, partly because fat is an insulator and impedes the passage of heat to the bacteria therein, and partly due to the need to fully inactivate an enzyme in the milk called lipase, which would otherwise break down the fat in high-fat products like cream or butter made from that cream, potentially giving rise to rancid off-flavors and aromas. However, lipase is not easy to assay in milk, and so an enzyme called lactoperoxidase (peroxidase for short) is typically used as the indicator, as it is both relatively easy to measure and a little more heat resistant than alkaline phosphatase, needing to be heated to a temperature around 78°C for 15–30 seconds (typical cream pasteurization conditions) to be inactivated.

In fact, sometimes milk processors assay both alkaline phosphatase and lactoperoxidase to fine-tune their process, by ensuring that the minimum conditions required to give a safe product (as indicated by a negative phosphatase test) have been reached, but yet not heating the milk excessively (the milk retains positive peroxidase activity). In this way, the two tests are used sometimes to "bracket" the exact heat treatment conditions, showing how useful enzymes are in validating and optimizing food heating treatments.

## Extending the shelf-life of milk

One of the greatest limitations to the shelf-life of pasteurized milk, as well as one of the potential health hazards that may be associated with it, is the presence of spore-forming bacteria such as *Bacillus* spp. So, to extend the shelf-life and/or to ensure a greater degree of safety, it is necessary to remove these from the milk.

In considering processes that have a greater antimicrobial efficacy than pasteurization, a key question is whether the objective is to reduce these spore-formers to a greater extent than heat treatment can achieve, knowing that the shelf-life (and risk) are directly correlated with their level, or else eliminate them entirely. If the latter is the goal, then this should result in a product that is effectively sterile, and so can be stored at ambient temperature. On the other hand, if spores are reduced in number rather than eliminated, the product will still require refrigeration, but will have a much longer shelf-life. There is really only one reliable method for eliminating spore-forming bacteria, at present, which is to heat them under conditions that will inactivate a typical level of spores that might be present, allowing for a comfortable margin of safety (as even one surviving spore that germinates could cause spoilage or safety problems).

On the other hand, if the goal is to physically remove spores, they could be considered as small solid particles present in the milk, like sand in water, for example. To remove sand suspended in water, there are two physical options: centrifugation or filtration. If applied to spores, these processes would remove (but not kill) perhaps 90%–99% of the spores. In the case of milk, the relevant operations are called bactofugation (bacterial centrifugation) and microfiltration, respectively.

Bactofugation uses centrifugal separators operating on similar principles to the separators discussed earlier for separating solids from milk, or cream from milk (called disk stack centrifuges). In this case, their design is optimized to remove the small, dense bacterial spores, in a small volume of milk that is either discarded or heated intensely to a temperature sufficient to inactivate the spores. While this might seem to be contrary to the idea mentioned earlier of reducing the overall heat load applied to the milk (and hence the nutritional and sensory changes caused), in practical terms, only a small fraction of the total volume receives this heat treatment and so when blended later with the larger volume of milk that was heated much more mildly, the overall impact on the taste is much less than for milk that was all heated to high temperatures, such as UHT milk.

For microfiltration, milk is typically first separated into cream and skimmed milk, which usually results in 10 times more of the latter than the former. The skimmed milk is then filtered under pressure using a polymeric membrane with ultra-fine pores (diameters between 0.1 and 10 micrometers) that allow all the milk components (especially the casein micelles) to pass through but retains the bacterial cells, which are larger, in a small volume. This volume can be discarded or heat-treated to sterilize it and then remixed with the skim milk. The cream is not filtered because the fat globules are too large to pass through the filter pores, but is typically heated sufficiently to inactivate the spores and then mixed with the skim, so that the overall heat load applied to the milk is still low, and a product with a reduced numbers of spores and hence an extended shelf-life (ESL), but high nutritional and sensory quality, is produced. The principles of microfiltration are discussed in more detail in Chapter 10.

These processes do not completely eliminate spores, as UHT treatment does, but rather reduces them sufficiently to allow a longer shelf-life than pasteurized milk (perhaps 1–2 months rather than 2–3 weeks), but under refrigerated conditions. In some countries (such as the United Kingdom, Canada, and France), ESL milk is quite popular.

## Sterilization of milk

As previously described, the D-value is the time required at a particular temperature to kill 90% of a certain type of bacterium, or, in other words, to give any individual cell a 1 in 10 chance of surviving this process. Consider the chance of survival of a potentially very dangerous pathogen like *Clostridium botulinum*. If you were told there was a 1 in 10 chance of a sample of milk containing a very dangerous bacterium, would you taste it? Most people would not. What about 1 in 100, or 1 in 1,000 or 1 in 10,000? The key question here is what would be an acceptable risk that a particular process could leave behind a pathogen that, given the later storage conditions of the product (i.e., unrefrigerated), would essentially have the potential to grow and produce a lethal toxin. In this case, the acceptable risk is 1 in $10^{12}$, or 1 in 1,000,000,000,000, or 1 in a trillion. To put it another way, for a cell to survive a heat treatment that gives it only a 1 in $10^{12}$ chance of survival, there would need to have been at least $10^{12}$ cells present initially for one to survive.

Giving *Cl. botulinum* a 1 in $10^{12}$ chance of survival means essentially giving it no chance of survival whatsoever. The chance of winning one of the richest lotteries in the world (the US Powerball or Euromillions lotteries) by comparison is 1 in $10^9$, and so winning one of these is 1,000 times more likely.

How does this relate to heat treatment conditions used for milk? The D-value for spores of *Cl. botulinum* is 15 seconds at 121.1°C. This means that heating *Cl. botulinum* for 15 seconds at 121.1°C will give a spore a 1 in 10 chance of survival, heating for 30 seconds will result in a 1 in 100 chance, and so forth. To reach a 1 in $10^{12}$ chance then needs 12 times 15 seconds, or 3 minutes.

The D-value decreases as the temperature increases, so less time at a higher temperature is required to achieve the same degree of bacterial kill. In this case, it has been shown that at 138–140°C, what took 3 minutes at 121.1°C takes only 2–4 seconds. That is exactly the combination of time and temperature applied to milk in ultra-high temperature (UHT) treatment, which results in a milk with a very long shelf-life and that need not be refrigerated.

Of course, at such high temperatures, the nutritional and sensory quality (e.g., development of cooked flavor and brown color) of milk may be negatively affected, and so any heat treatment at these temperatures should be designed to ensure the milk spends as little time as possible under these conditions.

## UHT treatment technology

Interest in producing dairy products with a significantly longer shelf-life than that of pasteurized milk resulted in a number of developments in the 1940s and 1950s in the United States, United Kingdom, and France, that led to the introduction of aseptically packaged UHT milk in the 1960s.

The key to producing a high-quality product despite heating to a temperature as high as 135–140°C, is to minimize the time milk spends at these temperatures. The best way to do this is through the rapid addition and subsequent removal of steam at these high temperatures. The principle of how this is done relates to the properties of water.

Water can exist as ice, water, or vapor, which are states of increasing energy and decreasing attraction of water molecules for each other. In ice, water molecules are locked into a regular crystalline arrangement, while in water they retain attraction but much more loosely, and in the vapor state the molecules have very little mutual attraction.

When water is heated, energy is added to the system, and, at certain temperatures, energy is used not to change its temperature but instead to change its state. At 0°C, ice melts to water, and at 100°C, water changes from a liquid to a vapor. These are called *phase transitions*, and absorb a great deal of energy (the specific requirements for energy are called *latent heats*).

When water moves from a more energetic to a less energetic state, it releases energy. When steam condenses to water, excess energy must be "dumped," and this energy release is called the latent heat of condensation.

One of the fastest ways to heat anything is to add highly energetic steam directly to it, for example by injecting hot steam directly into milk (as done by baristas when heating milk for cappuccinos or lattes). The rapid temperature change causes the steam to condense, and in the process, it releases a huge amount of energy (the latent heat). This is absorbed by the milk, increasing its temperature greatly, effectively instantaneously. In UHT processing, the milk can thus be brought up to a temperature around 138–140°C in a fraction of a second, at which temperature it can be held for the desired time (perhaps 3 seconds). At that point, however, the milk needs to be cooled, and the added water, which otherwise would dilute the milk, must be removed.

Fortunately, both things happen simultaneously when milk gives up the condensed steam, which can be induced by manipulating pressure, which has a major influence on phase transitions such as evaporation. Evaporation involves water molecules gaining enough energy to escape into the air, and pressure can be visualized as the opposing force of the atmosphere pushing down on the liquid. Increasing the pressure increases the resistance, requiring higher temperatures for water to boil; conversely, reducing the pressure decreases the boiling point. Under circumstances where it is desirable to prevent water from boiling, this can be achieved by operating at a high pressure whereas, in other circumstances (e.g., when concentrating milk) where it is desirable to make water removal easier, a vacuum is applied to allow rapid evaporation at a temperature lower than 100°C, which is less likely to induce undesirable changes in the milk.

In the case of UHT treatment, when steam is added the pressure is increased, so that the mixture doesn't boil. However, if the pressure is suddenly decreased (after the milk has been held for the desired period of time), rapid evaporation occurs, fueled by energy taken from the hot milk. By balancing the pressures carefully, it is possible to remove the exact amount of water added in the form of condensed steam, while cooling the milk in a fraction of a second. In practice, these changes in pressure are achieved through changes in the dimensions of the system through which the milk is flowing. Constraining the milk into a restricted volume increases its pressure, while allowing it to expand into a much larger volume results in a rapid pressure drop. This is the basis for what is called direct UHT treatment of milk.

In a more gentle form of direct UHT treatment, called infusion heating, the milk is introduced into a chamber through which it flows under gravity as a stream with by steam swirling around it. It absorbs energy from this steam, heats, is held, and cooled again as for the injection system. The basic principles of infusion and injection UHT systems are illustrated in Figure 5.9.

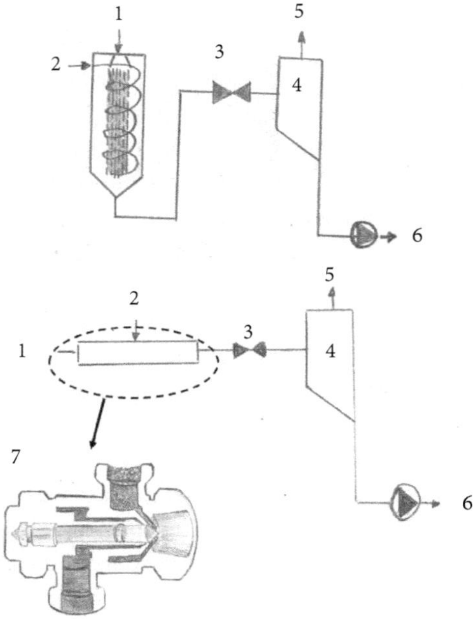


**Figure 5.9** Principle of operation of UHT systems showing (top) a direct infusion system, where milk (entering at 1) falls as a stream through a vertical chamber, to be heated by the steam entering at 2 and swirling through the chamber, followed by holding in the tube (3) between that chamber and (4) an expansion chamber at right, where a sudden pressure drop causes the condensed water to evaporate (5) and separate from the now cooled milk (6). In the injection system (bottom) the infusion chamber is replaced with an nozzle through which steam is injected (7), as shown schematically in the inset image (image thanks to Anne Cahalane).

Another type of UHT treatment process involves indirect heating, in which the heating medium (steam) and milk do not meet directly, but are kept separate by the walls of heat exchangers. For example, plate heat exchangers can be used, designed to operate at higher temperatures and pressures than those used for pasteurization. Other systems use tubular heat exchangers, based on concentric tubes that contain the hot and cold liquids in an inner tube or tubes while the heating or cooling medium flows in a surrounding larger tube, or shell. In a modification of the tubular system, the central tube contains a rotating shaft fitted with blades that scrape the inner wall as the liquid flows through it. Such scraped-surface heat exchangers are used for very viscous materials (e.g., ice cream mix) that do not mix easily and risk burning at the hot wall surface, which is prevented by the scraping motion.

## The shelf-life and quality of liquid milk products

Raw milk spoils quickly because of the large number of opportunities for growth in this nutrient-rich medium upon leaving the udder, and there are many different routes by which the milk can become unacceptable or unsafe. Of course, safety and spoilage are not the same, and milk could look unspoiled and have a normal taste and aroma but harbor pathogenic bacteria such as *Listeria monocytogenes* or *Salmonella* spp., which do not catalyze reactions that cause physical changes but represent a significant health risk if consumed.

The only process that is applied to raw milk is refrigeration, which will retard spoilage but will not eradicate pathogens, some of which (such as Listeria) are psychrotrophs. For these reasons, raw milk presents hazards, and its consumption is discouraged in most countries except under very limited conditions (which demand maintenance of a cold chain of storage and limit the usable shelf-life). The vast majority of milk intended for consumption is subjected to the types of heat treatment discussed above.

In summary, the goal of pasteurization is to inactivate all pathogens that might be expected to be present in raw milk. The temperature reached in pasteurization does not kill spores, but these are largely controlled (and unable to germinate and grow into an active and hazardous state) by keeping the milk refrigerated.

How then does pasteurized milk spoil? One key assumption underpinning the safety and quality of refrigerated, pasteurized milk, is the use of effective packaging that prevents recontamination of the milk. If a spoiled milk sample is found to contain bacteria that should have been eliminated by pasteurization, and the pasteurizer records show that the required temperature was attained during processing of that batch, then it can be concluded that the bacteria entered subsequent to heating. At one time, such post-pasteurization contamination (PPC) was relatively common (particularly before the principles of hygienic packaging and post-heating handling were fully understood) and represented a significant threat to the quality of pasteurized milk. The use of modern paperboard packaging systems such as Tetra Paks (first introduced in Sweden in 1951) has largely resolved this problem. PPC can also occur in consumers' homes if packages are left open.

In the absence of PPC, the main cause of spoilage is often the presence of a bacterium called *Bacillus cereus*, which has the unusual property of being able to survive both heating and refrigeration. It is spore-forming, so can survive heating in a dormant state, and it is a psychrotroph, meaning it can grow at low temperatures. It also produces enzymes that act on proteins in such a way as to destabilize and coagulate them. This can result in "bitty" milk, where white flecks of destabilized protein appear in the milk, while more advanced spoilage can lead to large coagula and separated whey in the package.

High initial levels of bacteria such as lactic acid bacteria in the raw milk can also lead to spoilage through acidification and ensuing protein destabilization, giving a yoghurt-like gel and aroma in the spoiled milk. If solid particles or gels appear in milk, this can usually be linked to proteins becoming destabilized either by acid or bacterial enzymes. Occasionally, other bacteria, e.g., *Alcaligenes* species, can cause spoilage through the production of extracellular polysaccharides (long, complex sugar molecules), resulting in ropy strings or threads in the milk.

The key variable controlling the rate of spoilage of pasteurized milk during storage is temperature, and even a few degrees can have a major impact on the rate at which milk will spoil. Extending the shelf-life of milk requires achieving greater reductions in bacterial populations, especially the heat-resistant spore-formers. One simple way to do this is to apply a higher heating temperature—ultra- or super-pasteurized milk may be heated at a temperature up to 120°C—with the heating time being shorter as temperature increases. However, this can cause changes in flavor through the introduction of cooked notes.

The other way to extend the refrigerated shelf-life of milk is to treat the spores as physical entities and remove them by suitable processes, such as microfiltration or bactofugation as discussed earlier. In such cases, elimination of all spores is not achieved, but every reduction of spore numbers by a factor of 10 or 100 extends shelf-life. The most effective of the ESL technologies is probably higher temperature pasteurization, followed by microfiltration and then bactofugation, which can add 6, 4, and 2 weeks, respectively, of refrigerated shelf-life to that achieved by normal HTST pasteurization.

To achieve much longer shelf-lives and avoid the need for refrigeration entirely requires not just reducing the number of spores but eliminating them completely, which can be achieved only by the severe heat treatment of UHT or retort sterilization. This results in a convenient milk product, but with inevitable compromises on flavor, color, aroma, and perceived nutritional quality.

This leads to an interesting trade-off between convenience and quality. Will consumers prefer a product with high sensory quality that has the inconvenience of needing to be kept in the fridge and spoils in a couple of weeks or one that is stable for months and can be stored anywhere? The answer to that question depends greatly on factors such as culture, geography, and the main uses for the milk. In some countries (such as Ireland, the United Kingdom, and parts of the United States), milk is very much a beverage for which people acquire a taste in childhood and which is regularly consumed even in adulthood. In such cases, flavor is one of the most important attributes of milk, and the cooked taste of UHT milk is less popular. UHT milk is, in such countries, found more frequently in small sachets or pots in hotel rooms or on trains or airplanes, where it is intended to be added to coffee or tea, where its stronger taste is not noticed.

In other countries with warmer climates, such as southern Europe, ambient temperatures make maintaining a cold chain more difficult, and UHT milk might be preferred. Interestingly, in countries that are not so warm (such as in Scandinavia) but do not have a tradition of consuming milk as a beverage, UHT milk is also seen as a more convenient product.

Wherever UHT milk is produced, it is typically sterile. But how then does it spoil? In terms of possible microbiological causes, some bacteria can actually withstand heating to 138–142°C, such as *Bacillus sporothermodurans*, but they are so adapted to high-temperature life that, even if they survive the heating, they will not grow unless the UHT milk is stored around 40°C, and so in most cases they do not present a problem.

Why then, in the absence of such bacteria, does UHT milk have any limitation to its shelf-life? The end of shelf-life for such a product, when it might be deemed to be "spoiled" and no longer acceptable to consumers, is usually when, on opening the container, some or all of the liquid milk has turned into either a very thick custard-like liquid or a semi-solid gel. In other cases, the milk might show floating pieces of white gel-like material. The term for these phenomena is "age gelation" of UHT milk.

As in the case of yoghurt and cheese, the presence of a solid gel in any dairy product is usually indicative of the destabilization of proteins through enzyme action or acid production. It is very rare for acidification to occur in UHT milk, unless some form of recontamination occurs, which leads to the likelihood of enzyme activity being responsible.

This is indeed the case, and one form of destabilization of UHT milk results from the action of the milk enzyme plasmin, which enters milk from the animal's blood and can break down the caseins in milk. Plasmin is very heat stable, and some activity can survive the intense heating of UHT processing and then act very slowly during storage. Plasmin does not form three-dimensional casein gels but rather is likely to give rise to the dispersed bitty gels mentioned earlier. The main strategy used to avoid the problem in UHT milk is to exploit the fact that the enzyme is actually easier to inactivate when heated at temperatures around 80°C than at higher temperatures (such as 138–140°C). Holding milk at this lower temperature (called preheating) for a few minutes before the main UHT treatment step gives a more stable product, due to more effective plasmin inactivation.

The other form of enzymatic destabilization is due to enzymes produced by psychrotrophic bacteria like *Pseudomonas* spp., that are able to grow at refrigeration temperatures but are readily killed by heat, even at temperatures around 65°C. However, they secrete proteases and lipases into the milk, which are very heat resistant, once they grow to high numbers. While the bacteria are inactivated by the heat, their enzymes can thus at least partially withstand UHT treatment and then act slowly during storage; it could be said that the milk is spoiled by the "ghosts" of long-dead bacteria.

The modes of action of these bacterial proteases are actually quite similar to that of rennet (chymosin) in that they preferentially act on κ-casein, the critical protein on the surface of casein micelles that maintains their stability. If any rennet-like enzyme acts on casein, it destabilizes the micelles and they aggregate, in the presence of a sufficient level of calcium, into a three-dimensional gel. If a very low level of an enzyme with a similar action is present in UHT milk and is left to act undisturbed for very long periods of time (months), then a cheese-like gel can form that can span the container, and can even expel whey, while mechanical damage due to moving or disturbing the container will result in substantial gels being present in the milk (an example of this is shown in Figure 5.10).

This is a very common form of destabilization of UHT milk; many common bacterial species that contaminate raw milk are known to produce these heat-resistant enzymes. In fact, some spoilage of pasteurized milk can also arise due to a similar mechanism acting over a shorter period of time. The key factor in controlling the problem is that enzyme production requires bacteria to grow to quite high numbers. The likelihood that these bacteria contaminate the milk from the environment is high, and so it is critical to process the milk before there is a chance for extensive growth, by either holding the milk at the lowest non-freezing temperature possible (their growth being slower at 2°C than at 4°C) or processing it as quickly as possible after milking (ideally within one day). In cases where logistics do not permit such rapid UHT processing, milk is sometimes heated under quite mild conditions (such as 65°C for 15 seconds, a process called thermization) to inactivate not the enzymes but the bacteria, before they have a chance to grow and produce the enzymes. Overall, in the case of problems due to bacterial enzymes, the principle is that prevention is better than cure, as elimination of the enzymes is essentially impossible, so the best strategy is to avoid their presence to begin with.

In terms of the shelf-life and quality of UHT milk, it must also be remembered that there are two main families of such products, produced by direct or indirect treatment. As described earlier, the extremely rapid heating and cooling rates with direct heating lead to the milk spending less time at very high temperatures, leading in turn to many fewer undesirable changes, such as cooked flavor, odor, color, protein denaturation, and nutrient destruction, and giving a more desirable (if more expensive) product. However, the lower heat load in directly heated milk results in less inactivation of enzymes, and so the shelf-life of the higher quality product can actually be shorter than that of the more severely heated indirectly treated product. This is another example of how processors trade off different attributes of a product that their consumers will value, such as quality versus shelf-life, noting that shorter shelf-life products

**Figure 5.10** A sample of gelled UHT milk, showing the settled curd-like gel and clear whey-like upper layer.

have less potential for export to distant markets where shipping times to reach their destination can represent a significant proportion of the stable life of the product.

In terms of retort-sterilized milk products, such as canned evaporated milk, gelation or thickening can also occur during storage, for similar reasons to UHT milk, but the overall heating applied is usually so severe that enzyme inactivation is much greater and, consequently, their shelf-life is very long.

In fact, for such products, a greater problem is likely to be coagulation during the process of heating. Heating at a temperature above 110°C results in a wide range of changes in milk. Whey proteins unfold, denature and interact with other proteins to form structures that, if they grow too large, can lead to gel formation and coagulation. The unfolding of these proteins also exposes sulfur-containing amino acids that lead to the typical cooked aroma and flavor of heated milk. In parallel, lactose breaks down to a range of compounds, including acids, and interacts with proteins through the Maillard reaction, leading to development of a brown color. All these changes happen in UHT milk, but the short heating time used means that they occur to a lesser extent than when milk is heated for much longer times at a lower temperature, as in sterilized canned dairy products. The net result of many of these reactions can be coagulation in the can during heating.

The mechanism by which milk becomes destabilized on heating has been studied widely by placing milk samples into glass tubes and putting them in a bath of hot oil at around 140°C, in which they are rocked gently. The time until visible coagulation occurs is recorded, which is called the heat coagulation time (HCT). This is frequently tested for a number of sub-samples of the milk, which have been adjusted to different pH values (in the range 6.2–7.2, where the initial pH of milk is usually 6.6) and a profile of the impact of pH on the time to coagulate can tell a lot about the properties of the milk and its likelihood of coagulation on heating (and also gelation during storage of UHT milk).

While proteins are usually the key structure-forming agents when milk coagulates, calcium, which enables the formation of such structures by cross-linking proteins, is also important. For this reason, a common way of increasing the heat stability of milk, as mentioned earlier, is to add salts such as phosphates or citrates that bind calcium and prevent it from destabilizing the milk.

## Fortified and flavored milks

Not that long ago, "milk" was a relatively simple product category on supermarket shelves, including whole (full-fat) milk, skimmed (largely fat-free) milk, and something in-between (low-fat). Today, however, the shelves of many markets include a wide range of variants of fluid milk, including products fortified with additives including minerals, vitamins, protein, and even fish oils, products in which lactose has been hydrolyzed, and flavored milks.

The rationale for vitamin fortification of milk is clear in many countries, where high levels of milk consumption by a population represent an opportunity to address certain widespread nutritional deficiencies. An example of this might be vitamin D in countries where, especially during the winter, there is not enough exposure to sunlight to produce this vitamin. In the case of other fat-soluble vitamins, such as vitamin A, they may be added to reduced-fat milk to replace that which has been removed in the cream by skimming the milk.

In many cases, fortifying or modifying milk is relatively straightforward. Adding vitamins or minerals can be done simply when levels of addition are very low and the components concerned are soluble. It may be a little more complicated for vitamins that are fat-soluble, but pre-mixes of vitamins (typically synthesized chemically) in concentrated oil- or water-based form are available, which can be added to milk to form a stable emulsion. If the components being added are heat-sensitive, they may be added following pasteurization.

In some countries, milk is also fortified with omega-3 fatty acids (FAs), which are long-chain, polyunsaturated fatty acids that include alpha-linolenic acid (ALA), eicosapentaenoic acid (EPA) and docosahexanoic acid (DHA). These health-promoting fatty acids have anti-inflammatory and anti-coagulant properties and are important for brain and eye development and heart health. Fish and fish oils are the most common sources of these components, but low consumption of these in some people's diet has led to the idea of including them in a more commonly consumed product like milk. The levels of these in milk can actually be enhanced in a more natural manner by increasing the levels of fish oil in the cows' feed, which leads to higher levels of the relevant acids being found in their milk, but another solution is to add the FAs directly to milk, typically after final heat treatment, due to their heat sensitivity.

One negative consequence of this is that such FAs are susceptible to oxidation, chemical reactions that can result in the development of off-flavors and odors (and perceptible rancidity) during storage. However, this can usually be avoided by modern packaging, which excludes the key catalysts for oxidation - light, and oxygen.

In terms of flavor addition to milk, some, such as vanilla, strawberry, or banana, are relatively easy to add in the form of a concentrated natural extract or in synthetic form, but some are more complex. Specifically, the production of chocolate milk has long been a challenge, as giving a rich chocolate flavor requires suspension of cocoa particles which, being more dense than the milk, tend to settle under gravity during storage, leading to a dense solid layer or sediment at the bottom of a container. To prevent this from happening requires the presence of a structure which somehow entraps the cocoa particles in suspension and prevents them from sedimenting, but yet does not affect the viscosity and mouthfeel of the milk. This sort of "fluid gel" can be produced by adding the polysaccharide carrageenan, obtained from seaweed, to the milk. This forms a very weak network that keeps the product liquid but yet has just enough gel structure to prevent the cocoa particles from sedimenting.

For those who suffer from lactose intolerance, which includes a large proportion of the global population (as discussed in Chapter 2), lactose-free milk products are attractive. It is relatively easy to produce the enzyme, β-galactosidase (which hydrolyzes lactose into glucose and galactose) commercially, so its application in the production of such products has been a viable option for many years. There are several ways in which lactose-free milk can be produced. One is to filter the milk through membranes (using technologies which we will discuss in detail in Chapter 9) to remove most of the lactose while retaining most of the other milk components, and then adding a small amount of β-galactosidase to hydrolyze the remainder. The enzyme can also be added to a tank of milk and allowed to react before standardization, homogenization, and pasteurization (which inactivates the enzyme) or else be added directly to the milk at the packaging stage, taking advantage of the shelf-life of the product to allow the enzyme sufficient time, even at refrigeration temperatures, to hydrolyze the lactose present before consumption.

Another common trend in many countries is market milk products with higher levels of protein than typically found in milk. This is usually achieved through the addition of dried forms of semi-purified milk proteins, such as Milk Protein Concentrates, the production of which will be discussed in Chapter 10.

One unusual category of milk products is milk that is obtained from cows milked while it is dark, which is reported to contain higher levels of a hormone called melatonin, which is involved in regulating sleep cycles. Such milk is sometimes marketed as a sleep aid for children or jet-lagged adults and is reported to assist in getting a sound night's rest.

# 6
# Cheese

## The origins of cheese-making

Cheese is the generic name for a large group of fermented dairy products, with a great diversity of flavors, textures, and shapes, produced throughout the world. Cheese was quite possibly produced by a "happy accident" thousands of years ago in the Fertile Crescent (the region around modern-day Iraq) when nomadic people used animal stomachs for transportation of milk. The tissue of such stomachs contains enzymes (especially that now called chymosin) that can coagulate milk. During storage, this would have been extracted and, at the warm ambient temperatures, caused the milk to coagulate and separate into a mixture of curds and whey. The whey could have been consumed as a refreshing drink, while the curds (especially if mixed with salt) was a way of preserving the goodness of milk for transport and later consumption.

At that time, two prerequisites for cheese-making, the availability of surplus goat and sheep milks (cows were not domesticated until much later) and of pottery utensils in which milk could be collected, stored, and coagulated, existed. Prehistoric pottery was unglazed and porous and adsorbed animal fats, residues of which could remain in the pottery for thousands of years. The earliest direct evidence of cheese-making comes from the discovery in 2012 in Poland of vessels, over 7,000 years old, that resembled cheese strainers in which traces of milk fat were identified.

Milk is also a rich source of nutrients for bacteria that contaminate milk and grow well at ambient conditions. Some of these, especially lactic acid bacteria (LAB), can use the milk sugar, lactose, as a source of energy, producing lactic acid and, consequently, reducing the pH of milk from its initial pH of 6.6 to ~4.6. The caseins coagulate to form a gel, within which the fat and aqueous phases of milk are incorporated. An acid-induced milk gel is quite stable, if left undisturbed, and is consumed as such in a wide range of fermented milk products (see Chapter 7). However, if cut or broken, the gel separates into curds and whey. The shelf-life of the curds can be extended by dehydration and/or salting to yield acid-coagulated cheeses, e.g., cottage cheese, cream cheese, and quarg, which represent ~25% of worldwide cheese production; these are usually consumed fresh, without maturation.

*From Farm to Table.* Alan Kelly, Patrick Fox, and Tim Cogan, Oxford University Press.
 DOI: 10.1093/9780197581025.003.0006

Alternatively, milk may be coagulated by proteinases from bacteria, molds, plants, or animal tissues (collectively referred to as rennets), including those originally extracted from the animal stomach containers; these modify the casein system, causing it to coagulate in the presence of calcium. Rennet-coagulated curds have better ability to expel whey (known as syneretic properties) than acid-coagulated curds, which makes it possible to produce lower-moisture, and hence more stable, cheese. For this reason, rennet coagulation has become the principal method for the coagulation of milk for cheese manufacture, and ~75% of the cheese produced today around the world is produced using rennet.

During the storage of rennet-coagulated curds, various bacteria and, in some varieties, yeasts and molds, grow, and the enzymes from milk, rennet, and microorganisms act on the milk components in the curd, resulting in changes in the flavor, texture, and functionality of the curd. When controlled, this process is referred to as ripening (maturation), during which a great diversity of characteristic flavors (due to a very complex mixture of molecules, mostly produced in trace amounts) and textures (due mainly to proteolysis and changing interactions of minerals such as calcium with protein) develop. The principal traditional rennets used for the manufacture of long-ripened cheeses are extracts from the stomachs of young mammals, particularly calves, in which the principal enzyme is chymosin.

Cheese manufacture accompanied the spread of civilization through Egypt, Greece, and Rome and was described by many Roman writers, especially Columella (4BC–AD75), who gave quite a detailed description of cheese-making in his book *De Re Rustica*. Columella's recipe for cheese involved coagulation of milk with rennet from kid goats or lambs, or, interestingly, thistle, wild saffron, or figs; draining the whey from the cut gel using cloths; and pressing and salting the resulting curds. Apparently, the cheese was consumed fresh and, when necessary, additional preservation was obtained by salting and smoking.

The daily ration of Roman soldiers included ~20 g of cheese and, as the Roman Empire grew, cheese-making spread over wide geographical areas. After the fall of the Roman Empire, the migration of people throughout Europe further promoted the spread of cheese manufacture. Subsequently, monasteries and feudal estates were key contributors to the development of cheese-making and the evolution of cheese varieties during the Middle Ages. Many cheese varieties made today, including Wensleydale, Port Salut, Saint Paulin, Munster, Trappist, and Époisses were developed in monasteries. The self-contained nature of communities based around feudal estates or monasteries led to wide local variation in practice and traditions, which gave rise to the emergence of a wide variety of types of cheese.

The evolution of cheese varieties was also influenced by particular local circumstances, such as species or breed of dairy animals available, local vegetation, or an "accident" during the manufacture or storage of the cheese, perhaps the growth of molds and/or other microorganisms. Those accidents that led to desirable changes in the quality of cheese were incorporated into the manufacturing protocols, and so cheese recipes and characteristics continued to evolve.

The later colonization by Europeans of other parts of the world, such as North and South America, Oceania, and Africa, introduced cheese-making to these regions, and cheese, mainly European varieties, sometimes modified to meet local conditions, has become an item of major economic importance in the Americas, Australia, and New Zealand.

The oldest known cheeses are those found around the necks of naturally mummified bodies in a cemetery in the Taklamakan Desert of northwest China in 2000, which were carbon-dated to have lived between 1600 and 1400BC (Figure 6.1). Analysis of protein extracts from these cheeses showed the presence

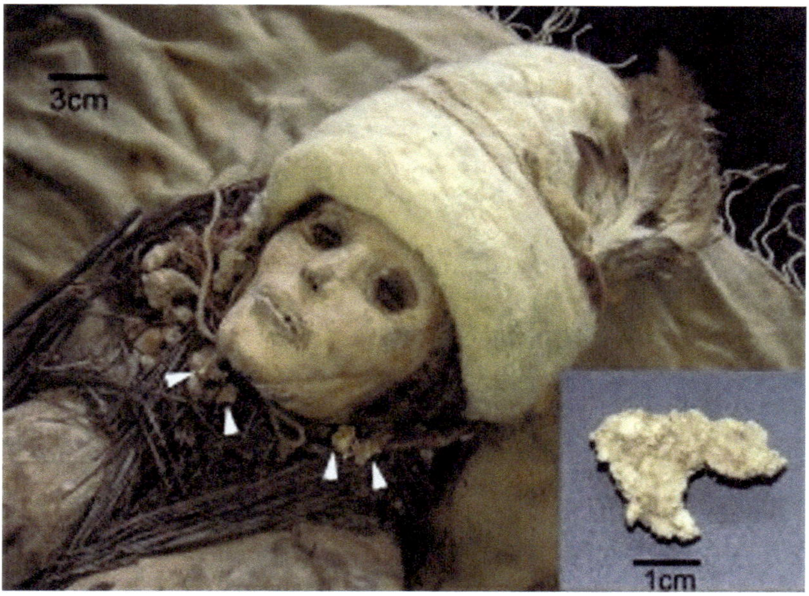


**Figure 6.1** Photograph of mummified body showing a cheese necklace and an insert showing an individual piece of cheese (from Yang et al. 2014. Proteomics evidence for kefir dairy in Early Bronze age China, *J. Arch. Sci.* 45, 178, with permission).

of the bacterium *Lactobacillus kefiranofaciens* and the yeast *Kluyveromyces marxianus*, suggesting that the cheese was made with kefir grains.[1] Another ancient cheese, dating to 1200BC, was found in the tomb of a pharaoh in 2014. Analysis of this cheese showed the presence of *Brucella melitensis*, the cause of brucellosis, and also indicated that the cheese was made from a mixture of bovine and sheep or goat milks.

Cheese-making remained a small-scale craft until relatively recently; now there are thousands of farm-scale cheese-makers and wide variation even within one general type of cheese. The first standardized method for making Cheddar cheese was developed in the United Kingdom in the mid-19th century; prior to that, it was produced in a number of sites around the village of Cheddar in Somerset. The first cheese factory in the United States was established near Rome, New York, in 1851, and the first in Britain at Longford, Derbyshire, in 1870.

Today, in spite of very considerable scientific and technological advances and intensive research, there is considerable inter- and intra-factory variation in the quality and characteristics of even well-defined varieties, due to the inherently complex biological nature of cheese production and ripening. The curds for many famous varieties of cheese, e.g., Parmigiano Reggiano and Grana Padano, are still produced on farms under the supervision of a producer consortium, and the cheeses are ripened in central facilities.

The current state of understanding of cheese science means that it is now feasible to systematically develop cheese variants with specified flavor, texture, and functional properties by controlling milk composition, standardization (i.e., adjustment of the fat and protein content, to give a specified fat:protein ratio in the cheese-milk) and other pre-treatments (e.g., bactofugation to remove bacterial spores). Careful selection of starter cultures, coagulant, and other ingredients, advances in manufacturing equipment, automation of most stages of manufacture, and better control of ripening conditions (packaging, time, temperature, humidity) have contributed to far more consistent cheese quality and facilitated the development of new varieties.

Examples of new cheese varieties developed in the 20th century include Jarlsberg, a Norwegian cheese with large eyes introduced in the 1960s; Maasdamer or Leerdamer, in the 1980s in the Netherlands to compete with Emmental, being cheaper and easier to make; and Dubliner, an Irish cheese developed in the 1990s due to a desire to diversify the number of varieties that could be produced in plants designed for Cheddar production.

There has been a marked resurgence in farmhouse cheese-making in recent years in many countries, in line with the increased demand for locally produced and artisanal food products; many of these cheeses could be regarded as new varieties.

## Cheese production and consumption

Cheese is the quintessential convenience food: it can be used as a major component of a meal, as a dessert, as a component of other foods or as a food ingredient, and it can be consumed without preparation or used in various cooking recipes.

About half of natural cheese is consumed directly as "table cheese," while a considerable amount is used as a food ingredient, e.g., Parmigiano Reggiano or Grana Padano on pasta products, mozzarella on pizzas, quarg in cheesecakes, and ricotta in ravioli. In addition, natural cheese is used in the production of a broad range of processed cheese products, which have a range of applications, especially as spreads, sandwich fillers, or food ingredients. Other cheese-based products include cheese powders and enzyme-modified cheese, both of which are becoming increasingly important as food ingredients.

World production of cheese in 2022 was ~22.2 million metric tonnes and has increased at a rate of ~2.2% per annum since 2005. While Europe and North America are the principal cheese-producing regions, Asia has the highest average growth rate at present.

Cheese consumption has increased consistently in most countries for which data are available; along with fermented milks, cheese is the principal growth product within the dairy sector. There are many reasons for this, including a positive dietary image, convenience, keeping quality, flexibility in use, and a great diversity of flavors and textures. The most rapid growth in cheese consumption in recent years has probably been as a food ingredient.

Some data on the production and consumption of cheese in various countries is shown in Tables 6.1 and 6.2.

## Classification of cheese

At least 1,500 varieties of natural cheese are produced today, and a number of attempts have been made to classify them into meaningful groups. Traditional classification schemes have been based principally on texture, which mainly reflects moisture content, i.e., extra-hard, hard, semi-hard/semi-soft, or soft. However, this scheme groups cheeses with widely different characteristics into one category; for example, Cheddar and Emmental are both classified as hard cheeses, although they have quite different textures and flavors due to different manufacturing procedures and ripening patterns.

Other ways to classify cheese in a more discriminating manner relate to the source of the milk (e.g., goat or sheep milk cheese), the method of coagulation or the principal ripening microorganisms (e.g., mold- or smear-ripened cheese).

**Table 6.1** Production of cheese in some countries in 2022

| Region/country | Tonnes produced (thousands) | Consumption per capita (kg) |
|---|---|---|
| United States | 6,217 | 17.3 |
| France | 1,856 | 26.5 |
| Germany | 2,361 | 24.1 |
| Italy | 1,374 | 21.8 |
| The Netherlands | 954 | 24.4 |
| United Kingdom | 504 | 11.8 |
| Canada | 499 | 14.5 |
| Brazil | 790 | 3.7 |
| Denmark | 455 | 28.9 |
| China | 283 | 0.1 |
| Ireland | 224 | 7.0 |

*Data sources*: Eurostat, US Department of Agriculture, Dairy Australia, Statista

**Table 6.2** Consumption of cheese in some countries and regions in 2022

| Region/country | Consumption per capita (kg) |
|---|---|
| France | 26.3 |
| Germany | 24.2 |
| Switzerland | 22.1 |
| The Netherlands | 19.4 |
| United States | 16.6 |
| Canada | 13.3 |
| Australia | 12.4 |
| United Kingdom | 9.6 |
| Ireland | 7.7 |
| China | 0.3 |

*Source*: World Population Review; Ireland for 2018 from Teagasc.

Based on the method of milk coagulation, natural cheeses may be divided into three super-families:

- rennet-coagulated cheeses, which includes most major cheese varieties;
- acid-coagulated cheeses, e.g., cottage, quarg, labneh, cream cheese;
- heat/acid-coagulated cheeses, e.g., ricotta, paneer, mascarpone.

A classification scheme based mainly on the means of coagulation is shown in Figure 6.2.

Acid and heat-coagulated cheeses are relatively minor varieties; they may be produced from whole milk, e.g., queso blanco (Central America) and paneer (India) or from rennet cheese whey or a blend of whey and skim milk and evolved as a means of recovering the nutritionally valuable whey proteins; they are frequently used as food ingredients. Important varieties include ricotta (Italy), anari (Cyprus), and manouri (Greece).

A further group of products, denoted "brown cheese" or brunost (e.g., Mysost), are produced in Norway. These are strictly not cheese, as they do not comply with the generic definition of cheese as a product prepared by coagulating the protein of milk but are rather made by boiling a mixture of milk, cream, and whey for several hours, during which a combination of evaporation and caramelization (through the Maillard reaction) yields a dark, sweet, solid product.

## Overview of cheese production

The production of most varieties of cheese involves a broadly similar protocol (Figure 6.3), various steps of which differ between varieties to give products with the desired characteristics. The principal features of the manufacture and characteristics of several cheese varieties will be described later in this chapter, but first an overview of the cheese-making process is described, so that the operations described later can be seen in an overall context.

### Selection of milk

The first major influence on the composition of cheese is the composition of the milk used, especially the concentrations of fat, protein, and calcium, and the pH. The composition of milk is influenced by several factors, as discussed in Chapters 1 and 2. Owing to major compositional abnormalities, milk from cows in the very early or very late stages of lactation and those suffering from mastitis should be excluded.

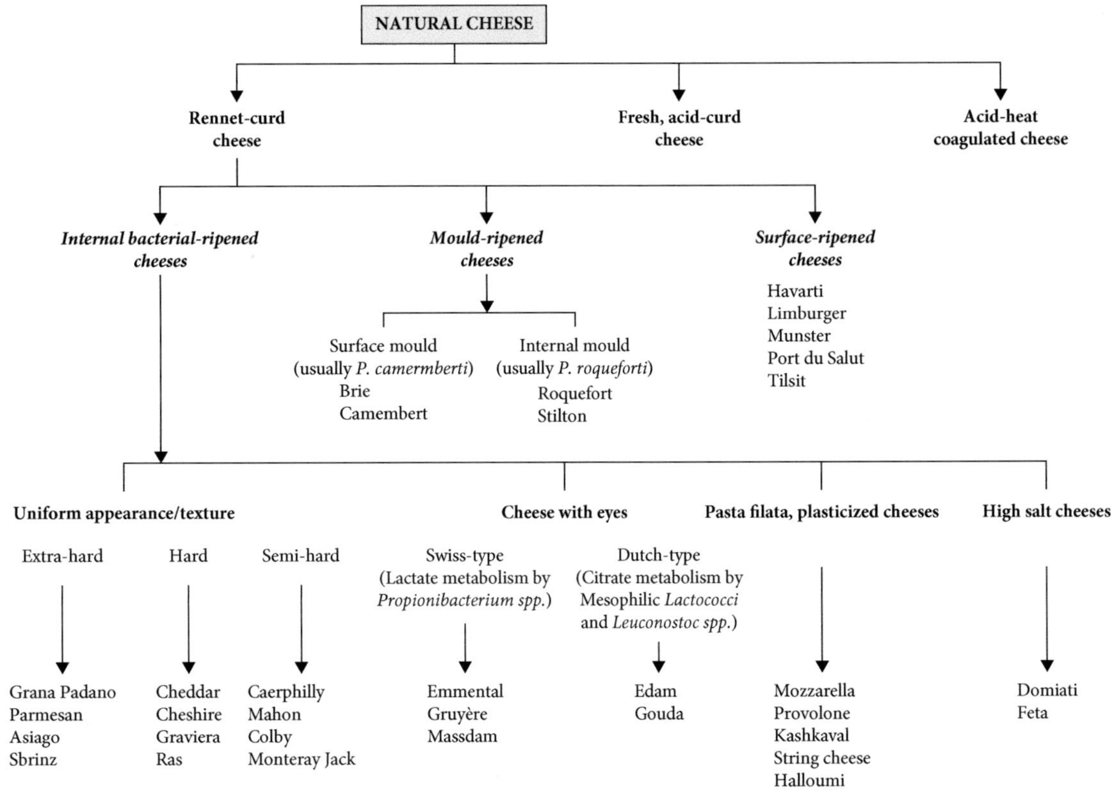


**Figure 6.2** A scheme for the classification of cheese (modified from Fox P.F. 2011. *Encyclopaedia of Dairy Sciences*. Vol. 1, 2nd edn.)

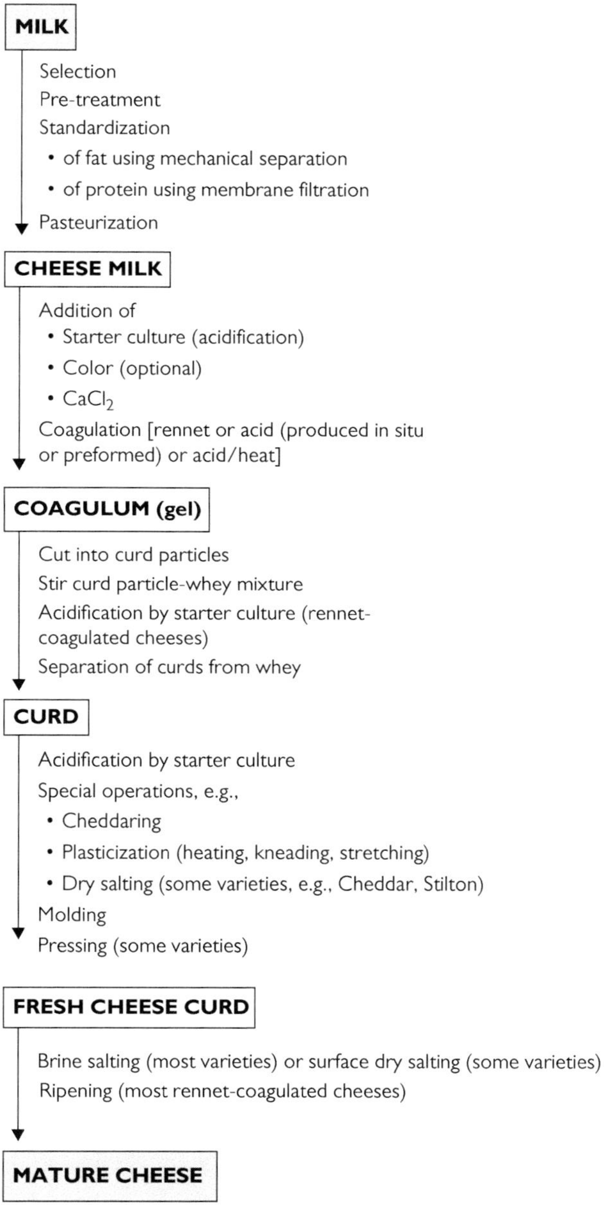


**Figure 6.3** General protocol for cheese manufacture (modified from Fox P.F. 2011. *Encyclopedia of Dairy Sciences*. Vol. 1, 2nd edn. Also, Fox et al., 2017. Fundamentals of Cheese Science Springer, New York).

A major influence on the characteristics of cheese is the species of animal from which the milk is obtained. Many world-famous cheeses are produced from sheep milk, e.g., Roquefort, feta, and Manchego, while traditional Italian mozzarella (Mozzarella di Bufala) is made from water-buffalo milk. There are very significant interspecies differences in the composition and properties of milk, which are reflected in the characteristics of the cheese produced therefrom. For example, goat milk contains specific strongly flavored short-chain fatty acids, which give cheese made from such milk a characteristic piquant aroma and flavor. There are also significant differences in milk composition between breeds of cattle, which influence cheese quality; the higher fat and protein content of the milk of Jersey cows compared to Holsteins leads to faster coagulation and higher yields of cheese made from the former.

The milk of yak and reindeer is used for local, small-scale cheese-making in some Asian and Scandinavian countries, respectively. The milks of camel, horse, and donkey yield a very weak or no gel and cannot be used for cheese production under typical cheese-making conditions using calf rennet. However, camel milk forms a relatively strong curd when coagulated with camel rennet.

Cheese milk should be free of chemical taints and free fatty acids, which cause off-flavors in the cheese, and antibiotics, which inhibit the growth of starter cultures. The milk used for cheese-making should be of good microbiological quality, especially for raw milk cheeses, as contaminating bacteria will be concentrated in the curd and may grow, causing defects or public health problems. For this reason, cheese milk is often pasteurized to render it free of pathogens and to reduce the numbers of spoilage bacteria. Raw milk may also be a source of non-starter lactic acid bacteria (NSLAB), which may play a role in cheese ripening, as will be discussed later.

## Standardization of milk composition

The composition of many types of cheese is prescribed with respect to moisture and fat-in-dry-matter, which, in effect, specifies that a certain fat:protein ratio is required. The moisture content of cheese, and hence the level of fat and protein, is determined mainly by the manufacturing protocol, but the fat:protein ratio is determined mainly by the fat:casein ratio in the milk, as the whey proteins are largely removed during cheese-making. The fat:protein ratio of cheese-milk can be modified by removing some fat by natural creaming (e.g. by gravity separation in the manufacture of Parmigiano Reggiano) or centrifugation, or adding, as appropriate, skimmed milk, cream, milk powder, evaporated milk, or ultrafiltration (UF) retentate.

Such additions may also increase the total solids content of the milk and hence increase the yield of cheese curd per unit volume of milk. However, the addition of high levels of skim milk powder (>1.0%) to standardize the fat content is generally undesirable because it results in a higher lactose content in the cheese milk, which can lead to a high concentration of lactic acid and/or residual lactose in the cheese. Excess lactic acid causes a low pH, and/or associated textural, flavor, and functional defects (e.g., crumbly texture, acid and bitter flavor development, excessive browning, and blistering on baking), while unfermented lactose may promote the growth of spoilage bacteria and reduce the acceptability of the product for lactose-intolerant people.

Calcium plays an essential role in the coagulation of milk by rennet; to overcome variations in the level of calcium in milk, it is common practice to add calcium chloride (e.g. 0.01%, w/w) to cheese milk.

The pH of the milk is also critical. In the past, it was often common practice to add the starter culture to the milk 30–60 minutes before adding the rennet, in a step designed to allow the culture to rapidly enter the exponential phase of growth and also favor rennet coagulation by slightly decreasing the milk pH (this is sometimes referred to as ripening but should not be confused with the later ripening of the cheese). However, highly concentrated direct-to-vat set starter concentrates (DVS) used today are added at much lower levels and have little or no direct acidifying effect, so ripening is not necessary.

## Application of membrane filtration in cheese-making

Since cheese manufacture is essentially a dehydration process, it seems obvious that membrane processes would have applications, not only for standardizing cheese milk with respect to fat and casein, but also for the preparation of a concentrate with a composition similar to that of the finished cheese, referred to as "pre-cheese." Standardization of cheese milk by adding UF concentrate (retentate) is now common, but the manufacture of pre-cheese has been successful commercially for only some varieties, most notably soft/semi-soft cheeses, such as UF feta, quarg, and Camembert.

A recent development is the manufacture of cheese or cheese-like products from reconstituted high-protein dairy powders, e.g., milk protein concentrate micellar casein, whey protein isolate, and/or whey protein concentrate prepared using membrane filtration (UF and/or microfiltration) (see Chapter 10). The reconstituted powders, together with other ingredients, including anhydrous butter oil, whey powder, and salt are blended in the desired sequence and heated to prepare a pre-cheese with the same total solids content as the final cheese. This is converted to cheese by adding a starter culture, food-grade acid, or an acidogen like glucono-δ-lactone (GDL), which breaks down to gluconic acid in milk. In contrast to conventional cheese manufacture, this approach involves concentration prior to gelation, as opposed to gelation followed by protein dehydration and concentration. Some advantages of this new approach include the near-elimination of whey expression during cheese-making; the potential to form composite gel systems by adding non-dairy ingredients such as gums; and the addition and retention of added materials such as bio-functional ingredients (e.g., prebiotic cultures), taste and aroma compounds, colors, and enzymes. This process is now used commercially for the manufacture of feta- and labneh-like cheese and high-protein dairy-based snacks.

## Heat treatment of milk

Traditionally, cheese was made from raw milk, a practice that was almost universal until the 1940s, and is still practiced for some varieties today. Raw milk cheese develops a more intense flavor than that produced from pasteurized milk, due to the greater contribution of heat-sensitive bacteria and enzymes from the milk (e.g., lipase) to ripening. Producing cheese from raw milk poses a public health risk, arising from pathogens potentially present in the raw milk; this is likely to be greatest in high-moisture varieties ripened for a short period (e.g., Camembert) and is greatly reduced in low-moisture varieties that are ripened for a long time, which are, in some countries, the main types of cheese for which the use of raw milk is permitted. Although a considerable amount of cheese is still produced from raw milk, especially in southern Europe (including such famous varieties as Swiss Emmental, Gruyère de Comté, Parmigiano Reggiano, and Grana Padano), pasteurized milk is now generally used.

When cheese was produced from very fresh milk on farms or in small local factories, the growth of contaminating microorganisms was minimal but, as cheese factories became larger, storage of milk for longer periods became necessary, and hence the microbiological quality of the milk became more problematic. Such storage conditions permit the growth of psychrotrophic bacteria, especially *Pseudomonas* spp., which secrete enzymes that affect the quality of cheese by breaking down the proteins and lipids present. Heat treatment of milk at a sub-pasteurization temperature, e. g., 65°C for 15 seconds (referred to as thermization), as soon as the milk is received at the factory, is fairly widely practiced in some countries (e.g. the Netherlands) to extend the storage period by inactivating these types of bacteria.

There are some alternatives to pasteurization for reducing the number of microorganisms in milk, such as bactofugation to remove *Clostridium* spores. Microfiltration (using small-pore membranes as discussed in Chapter 10 for separation of milk proteins) is very effective at removing bacteria and spores from milk but is not yet widely practiced in the cheese industry.

## Addition of cheese color

The principal pigments in milk are carotenoids, which are obtained from the animal's diet, especially where the animal is grass-fed. Cattle transfer carotenoids into milk directly and, therefore, bovine milk fat and high-fat products produced from it, including butter and cheese, are yellow to an extent dependent on the carotenoid content of the animal's diet, while cheeses made from sheep, goat or buffalo milk are very pale. A yellow color is often portrayed as an indicator of quality in the marketing of products such as butter produced in countries where milk production is mainly from grass-fed cows, e.g., Ireland and New Zealand.

Some consumers value highly colored cheese, which is usually achieved by adding annatto, extracted from the seeds of *Bixa orellana*, a shrub native to Brazil. Annatto contains two pigments, bixin, which is oil-soluble and norbixin, which is water-soluble. Oil-soluble annatto, rich in bixin, is used to color high-fat dairy products such as butter, whereas water-soluble annatto preparations, which contain mainly norbixin, are easily dispersed in milk and are used in the manufacture of cheeses such as red Leicester, red Cheddar, and double Gloucester. Many consumers believe that a deeper red color in cheese reflects a stronger flavor, although this is not the case as the added pigments are flavorless.

One issue associated with adding colors to cheese-milk is that a significant proportion (10%–20%) of water-soluble annatto is lost in the whey and gives color to whey-based ingredients, which may limit their applications. This has become a concern as the options for producing high-value ingredients from whey has increased. Consequently, various approaches have been explored to reduce the impact of annatto on whey-based ingredients, including the treatment of whey with a bleaching agent (e.g. hydrogen peroxide, which is allowed in the United States but not Europe) or the use of colorants not containing norbixin, e.g., paprika and/or β-carotene.

## Conversion of milk to cheese curd

After the milk has been standardized and pasteurized, it is transferred to vats (or kettles), which vary in shape (hemi-spherical, rectangular, or cylindrical, vertical or horizontal, open or closed), and in capacity (from a few hundred liters to 30,000 liters or more), where it is converted to cheese curd by a process that involves three basic operations: acidification, coagulation, and dehydration.

## Acidification

Acidification is a key event in cheese production and is usually achieved through the production of lactic acid by the starter cultures. Originally, the indigenous milk microflora performed this function but, since this was highly variable, starter cultures were introduced about 150 years ago and since then have been improved progressively and refined (see Chapter 4). Today they are used for most cheese varieties. The acidification of curd for some artisanal cheese varieties still relies on the indigenous raw milk microflora, which can lead to variability in cheese pH and shelf-life.

Direct acidification, by adding an acid (e.g. acetic, lactic, or citric acids) or an acidogen, is an alternative to biological acidification and is used commercially to some extent in the manufacture of cottage, quarg, feta, and mozzarella cheese. Direct acidification is more controllable than biological acidification, but enzymes from starter bacteria are essential in cheese ripening, and hence chemical acidification is used only for cheese varieties in which texture and/or functionality are more important than flavor.

The rate of acidification—or pH decrease—after starter addition depends on the amount and type of starter added and on the temperature profile of the curd. The final pH of the curd for most rennet-coagulated cheeses is 5.0–5.3, but the pH of acid-coagulated varieties (e.g. cottage, quarg, and cream) and some soft rennet-coagulated varieties (e.g. Camembert and Brie) is ~4.6. The time required to reach the final target pH ranges from 4–6 hours for cheddar and cottage cheese to 10–12 hours for Dutch and Swiss types. The pH of cheese changes during ripening and increases significantly in some varieties, especially surface-ripened varieties.

Production of acid at the appropriate rate affects several aspects of cheese manufacture and is critical for the production of good-quality cheese. It affects the coagulation process in several ways. For example, rennet coagulates milk better under slightly acidic conditions, and both the strength of the gel formed (curd tension) and the rate at which it synereses (expels whey), which influence cheese yield and composition, are also affected by pH. Critically, the final moisture content of cheese curd regulates the growth of bacteria and the activity of enzymes in the cheese and consequently strongly influences the rate of ripening and the quality of cheese. Acid production also influences how much coagulant is retained in the curd, which influences the subsequent rate of protein breakdown during ripening and may affect cheese quality.

In addition, the solubilization of colloidal calcium phosphate associated with the casein micelles during cheese manufacture increases as the pH decreases, as a result of which properties such as meltability and stretchability of the cheese are modified.

Finally, as low pH inhibits the growth of many non-starter bacteria in cheese, including pathogenic and gas-producing microorganisms, the level of acidification and final pH reached have a major influence on its safety and shelf-life.

## Coagulation

Rennet-induced coagulation of milk is achieved by the addition of selected proteinases, called rennets; these attack the κ-casein, which plays a key role in stabilizing the casein micelles, as we saw in Chapter 2. They split the protein molecule at a very specific point, between an amino acid called phenylalanine at the 105th position and one called methionine at the 106th position, which causes the removal of the part of the molecule that is responsible for the stabilizing effect. Following this reaction, the casein micelles are destabilized and aggregate to form a gel when the milk is held under quiescent conditions without agitation or stirring.

A number of enzymes can act as rennets in cheese-making and have different advantages and disadvantages. The ideal coagulant for milk has the following characteristics:

1. It preferentially hydrolyzes the peptide bond mentioned above, with little hydrolysis of the other bonds during manufacture. The highly specific cleavage of a single peptide bond results in a high recovery of casein (≥ 95% of total casein) in the cheese.
2. It promotes the onset of milk gelation in 15–25 minutes and the formation of a clot (coagulum, gel) with a firmness suitable for cutting and stirring without shattering after 35–45 minutes.
3. It has a low thermal stability (~ 90% inactivation after heating at 68°C and pH 5.3 for 1 minute), which means it can be readily inactivated by cooking in some types of cheese, where its action during ripening is undesirable (e.g. Emmental), or in whey during later processing.
4. It results in good quality cheese with the desired flavor (free from defects such as bitterness), texture, and functionality.

For these reasons, gastric (stomach) proteinases (chymosin and pepsin) from calves, kid goats, or lambs have been used traditionally as rennets in cheese manufacture. These were for a long time prepared by extracting the dried (usually) or salted gastric tissue (called vells) with 10% NaCl and standardizing the activity of the extract, which contains both chymosin and another digestive enzyme, pepsin, at levels of ≥75% and ≤25%, respectively. The secretion of chymosin declines while that of pepsin increases as the animal ages, which is why calf rennet was traditionally favored. Animal rennets with a high proportion of chymosin are preferred, as they favor higher cheese yields, more controlled proteolysis during maturation, and an overall more desirable cheese quality (flavor, texture, and functionality). Lamb and kid rennets are used sometimes for the manufacture of cheeses around the Mediterranean, such as feta and Pecorino Romano.

Milk can be coagulated by other proteinases, from animals, plants (e.g. thistle and nettles), or microorganisms, but most are too proteolytic relative to their milk-clotting activity (and so do not meet the first criterion mentioned above). Consequently, they lead to rapid and excessive non-specific hydrolysis of caseins in the coagulum during cheese manufacture, loss of casein as peptides in the whey, and a reduction in cheese yield. Excessive proteolysis or incorrect specificity may also lead to defects in cheese flavor (e.g. bitterness) or an overly soft texture.

Expansion of cheese production and a shrinking supply of calf vells led to the introduction of rennet substitutes in the 1960s. The first successful rennet substitutes were proteinases derived from microorganisms: *Rhizormucor miehei* or *Rhizormucor pusillus* (e.g. Rennilase, Fromase) or *Cryphonectria parasitica* (e.g. Thermolase). However, microbial rennets tend to result in higher levels of protein in cheese whey, lower cheese yields, and higher proteolysis in cheese. Moreover, compared to bovine chymosin, *R. miehei* and R. *pusillus* proteinases are more heat-stable and *C. pusillus* proteinase less so, which affects inactivation during heating of the curd (e.g. during cooking or curd plasticization) and can lead to higher residual activity in the cheese, more proteolysis during storage, and alteration in texture-functionality/ripening time profile. Nevertheless, microbial coagulants give satisfactory curd formation and are used in niche applications (e.g. use of *Rhizomucor* rennet to enhance proteolysis in high-cook cheeses).

In the 1980s, the use of recombinant DNA technology led to the introduction of fermentation-produced chymosin (FPC) by genetically engineered *Aspergillus niger* var. *awamori* or *Kluyveromyces lactis*. FPCs (e.g. Chy-Max and Maxiren) are identical to calf chymosin and are the main coagulant used today. In addition, camel chymosin (which has an even higher level of milk-coagulating activity relative to other protein breakdown) has been produced by fermentation in recent years and is used in production of many cheese varieties.

The activity of coagulants for milk is expressed in international milk clotting units (IMCU): one unit of activity is defined as the amount capable of coagulating 10 ml of milk at pH 6.3 in 100 seconds. So-called single-strength rennet contains 200 IMCUs per ml (IDF Standard 157:2001) and 200 ml of such rennet is sufficient to coagulate 1,000 kg of milk in 30–40 minutes at 30–32°C, the typical temperature at which milk is renneted in most cheese varieties.

## Stirring and cooking of curds

Rennet- or acid-coagulated milk gels are quite stable under quiescent conditions, but if cut or broken, they expel whey. This is called syneresis and is a key step in making cheese, as it concentrates the fat and casein of milk by a factor of 6–12, depending on the variety. The rate and extent of syneresis greatly influence the final composition of the cheese and are thus key points of differentiation of cheese varieties. Syneresis is influenced by the concentrations of calcium and casein, pH, cooking temperature, level of stirring of the curd–whey mixture, and time.

The next steps in the cheese-making process differ between varieties (as will be discussed later in this chapter) but have the function of reducing the level of moisture in curd particles by inducing expulsion of whey to an extent depending on the final target moisture content of the cheese.

Usually, following cutting of the rennet-induced gel, the mixture of curds (which have typically consolidated into small pea-sized curd particles at this stage) and whey is stirred, gently at first and then progressively more robustly, for a period of time (perhaps 30 minutes) to allow collisions between curd particles to promote whey expulsion.

In many cases, this is followed by "cooking" or heating, which causes the curd to contract and expel more whey. The temperature to which the curds and whey are cooked depends on the variety, and can range from 38°C for Cheddar to 55°C for Emmental (the temperature reached has a significant influence on enzyme inactivation and starter culture survival and final cheese quality and characteristics). In some cases (e.g. Cheddar), cooking is done by introducing hot water or steam into the hollow vat walls, while in other cases the cooking may involve removal of some whey and replacement with hot water (some Dutch cheese varieties). The latter process also dilutes lactose and soluble mineral levels in the whey.

In-vat whey expulsion is undesirable where a high moisture cheese variety is being produced, and the extent and duration or stirring and cooking are minimal before recovery and consolidation of the curds, while for low-moisture hard varieties the process of stirring and cooking can take 60–90 minutes.

Once the desired degree of whey expulsion from curds has been reached, the curds and whey are separated, by scooping curds out in the case of some varieties, or by using mesh or other filters to retain curd while the whey is drained by gravity from the vat. While for some types of cheese the curds are then placed directly into molds for pressing, in the case of Cheddar cheese there is a further process called cheddaring in which piles of curd are created and allowed to stand warm in the vat, pressing under their own weight into a homogeneous mass, before being milled into finger-sized pieces (curd chips) for salting and pressing.

## Molding and pressing

Once the coagulum or curd has been formed and the correct pH for the variety of cheese being made has been reached, the curd is put into cylindrical or square molds and pressed. The pressure exerted during pressing can be simply the weight of the curd itself, as in Camembert cheese, or pressures of several hundred kg per $cm^2$ generated by a hydrolytic ram, as in hard cheeses. Pressing helps to release more whey from the curd and ensure that the right level of moisture is attained in the cheese prior to ripening.

Whey contains about 50% of the solids in milk (98% of the lactose, 25% of the protein, and about 10% of the fat). In the past, whey was regarded as essentially worthless, to be disposed of as cheaply as possible, but today is regarded as a valuable co-product of cheese manufacture that is converted into high-value protein-based ingredient powders, such as whey protein isolates/concentrates and whey powders, which are used in infant milk formulae, sports nutritional beverages, and as highly-functional ingredients in food formulations (Chapter 10). Lactose may also be crystallized from whey and used in various applications, e.g., to adjust the composition of infant formulae, as a filler or filler-binder in the manufacture of pharmaceutical tablets and capsules, or for the production of glucose-galactose syrups, ethanol, or lactic acid.

## Salting

Almost all varieties of cheese are salted at the end of curd manufacture, by a number of different approaches:

- mixing dry salt with curd chips, e.g., Cheddar and related varieties;
- submersion in brine, e.g., Gouda and Emmental;
- rubbing dry salt on the surface of the pressed cheese, e.g., blue and smear-ripened cheeses.

The final level of salt, which varies from about 2 to 10% (w/w) in the moisture phase of cheese, has a major influence on cheese ripening, quality, and safety. In dry-salted varieties, the level of salt in the cheese moisture phase rapidly reaches a value (5%–6%, w/w) which prevents the growth of starter bacteria; therefore, the pH of the curds for these varieties at salting must approximate the ultimate value (pH 5.1). However, most varieties are salted by immersing the molded/pressed cheese in brine or by application of dry salt on the cheese surface. As the diffusion of NaCl into the interior of the cheese is then relatively slow, there is ample time for the pH to decrease to around 5.0 before the concentration of salt becomes inhibitory throughout the cheese. The pH of the curd for most brine-salted cheese varieties is 6.2–6.5 at molding and pressing but decreases to 5.0–5.2 during or shortly after pressing.

The Greek cheese feta is an exception, since it is ripened in brine containing 6%–8% NaCl for 2–3 weeks at 16–18°C, during which the pH decreases to around 4.6 and the moisture to ~56%, after which it is transferred to a cold room at 4–6°C for periods up to a year. In a few varieties, e.g., Domiati, salt (8%–15%) is added to the milk before renneting, which has a major influence on acid development and the growth of the microorganisms.

## Cheese ripening

Acid-coagulated cheeses are typically consumed fresh at the end of curd preparation. Rennet-coagulated cheese may also be consumed as fresh curd, and a little is, but most of these varieties are ripened (matured) for a period ranging from about 3 weeks to more than 2 years; generally, the duration of ripening is inversely correlated with the moisture content of the cheese. Most varieties may be consumed at any stage of maturity, depending on the flavor preferences of consumers and economic factors. For example, mild Cheddar is typically 3–6 months old, while mature Cheddar may be up to a year old and extra-mature older still.

The unique characteristics of each type of cheese develop during ripening as a result of a complex set of biochemical reactions. The changes that occur, and hence the flavor, aroma, and texture of the mature cheese, are largely predetermined by the manufacturing process, especially by the levels of moisture and salt, pH, residual coagulant activity, and type of starter used and, in many cases, by the secondary microflora (added or adventitious). The specific ripening conditions, such as temperature, humidity, and whether the cheese is packaged or not during ripening, also have a significant influence. Some traditional varieties of cheese are ripened in very specific environments, including cellars, caves, and pits, that likely contribute to their unique characteristics.

The biochemical changes that occur during ripening are caused by one or more agents:

- coagulant;
- indigenous milk enzymes, especially proteinase and perhaps lipase (especially in raw milk cheese);
- starter lactic acid bacteria and their enzymes;
- NSLAB and their enzymes;

- secondary microorganisms (e.g. *Propionibacterium*, *Brevibacterium*, *Corynebacterium*, *Penicillium* spp. and yeast) and their enzymes.

The biochemical reactions during ripening may be divided into three principal groups:

- Catabolism (break down) of lactose, lactic acid, and, in some varieties, citric acid; these cause changes in cheese pH, flavor, texture, and functionality. In Emmental, Gruyère, Jarlsberg, and Maasdammer cheese, propionic acid bacteria catabolize lactic acid to propionic acid, acetic acid, and carbon dioxide, which causes the formation of the characteristic large eyes in these cheeses. In smear-ripened and surface mold-ripened cheeses, the lactic acid is catabolized extensively to water and carbon dioxide on the surface of the cheese and the pH increases to ~7.5. The increase in pH causes calcium phosphate to migrate from the interior of the cheese to the surface and to precipitate as a layer there. The increase in pH and the loss of calcium cause the texture of the cheese to soften and become more fluid during ripening.
- Lipolysis, mainly by added lipase (pre-gastric esterase) as in some Italian varieties, e.g., Parmigiano Reggiano and Grana Padano. A low level of lipolysis occurs in most varieties, catalyzed by lipases produced by starter bacteria and NSLAB. Considerably more lipolysis occurs in cheese made from raw milk, due to the action of the indigenous lipase of the milk, which is inactivated by pasteurization. Extensive lipolysis also occurs in blue-mold cheeses due to the action of lipase secreted by *Penicillium roqueforti*, which is added to the milk before coagulation; this mold also catabolizes the liberated fatty acids to methyl ketones, which dominate the flavor of blue cheeses.
- Proteolysis and modification of amino acids. These are the most complex and perhaps the most important reactions in many varieties, as they affect flavor, texture, and functionality (as protein is the key structure-forming component of cheese). In some varieties, proteolysis is relatively limited, and affects mainly texture and functionality, but the extent of protein breakdown is much more extensive in mold- and smear-ripened varieties. In Camembert, extensive transformation of amino acids to ammonia by enzymes secreted by the mold occurs at the surface, which contributes to the pH gradient from the surface to the center mentioned above.

The quality of cheese is determined by its flavor and aroma, texture, physical appearance (presence of eyes, growth of molds or pigmented bacteria), and physicochemical properties (firmness, viscosity, color, melting properties).

The relative importance of these characteristics varies in different varieties, and grading is usually done by experienced individuals, trained panels, or consumer panels.

In addition, chemical, microbiological, and/or physical analyses are frequently performed to characterize cheese ripening. For example, proteins in cheese, and their large breakdown products, can be separated under an electric current in a procedure called polyacrylamide gel electrophoresis, while small peptides can be separated by various types of chromatography. Volatile aroma compounds that contribute to cheese flavor can be separated by gas liquid chromatography and then identified by mass spectrometry.

Changes in texture may be monitored by various rheological methods that measure the deformation or flow of a material when subjected to mechanical stresses, which may also be used to study the melting characteristics of cheese. Microbiological analysis, including counts of particular species of microorganisms and bacteriophage in the milk and cheese, may also be carried out.

## Development of microorganisms in cheese during ripening

Two groups of lactic acid bacteria are found in cheese, the starter lactic acid bacteria (SLAB) and NSLAB. SLAB dominate the microbiology of all cheese types soon after manufacture, reaching maxima of $\sim 10^9$ cells per gram in the cheese within a few hours. In the first weeks after manufacture, these bacteria die off and their cells lyse, releasing enzymes that contribute to the ripening of the cheese. They are replaced by a secondary microflora of NSLAB, which multiply from low numbers ($<100$/g) in the fresh cheese to $10^8$–$10^9$/g during ripening. NSLAB mostly comprise lactobacilli, particularly, *Lb. casei*, *Lb. paracasei*, *Lb. curvatus*, and *Lb. plantarum*, and less frequently obligate heterofermentative species like *Lb. brevis* or *Lb. fermentum*. *Pediococcus* and *Enterococcus* spp. are also found (Figure 6.4). The source of NSLAB is unclear. They may be adventitious contaminants from the environment and/or milk processing equipment, and so the population in a particular type of cheese might be unique to the production environment and plant in which it is made. Some strains of NSLAB isolated from cheese withstand pasteurization, implying that raw milk may be their source, but other strains do not survive pasteurization, implying that post-pasteurization contamination of the milk is the source. The high numbers of NSLAB in the ripening cheese suggest that they must have some effect on the development of flavor, and local differences in NSLAB populations might be another reason for the great diversity of cheese flavors even for the same variety made in different locations.

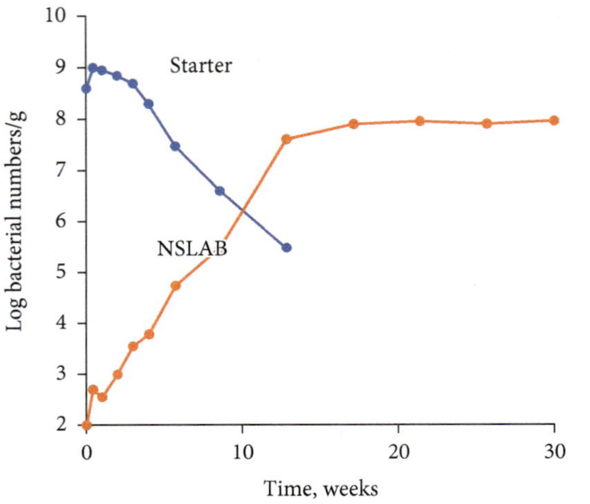


**Figure 6.4** Development of starter lactic acid bacteria (SLAB, blue) and non-starter lactic acid bacteria (NSLAB, red) in a typical Cheddar cheese during ripening (from Fox et al. 2017. *Fundamentals of Cheese Science*. 2nd edn. Springer, New York, with permission).

## Adjunct cultures

Much commercial hard cheese made today is considered to have a mild flavor. The probable reasons for this are low bacterial numbers in the milk when the cheese is made from raw milk, the killing effect of heat when pasteurization is practiced, and good hygiene in milk production and cheese factories. Because of this, various methods for enhancing flavor have been developed.

Traditionally, only mesophilic cultures were used in the production of Cheddar and other low-temperature cooked cheeses. However, in recent years, thermophilic starters, particularly *Sc. thermophilus*, are also being used, together with the mesophilic starters, to improve the flavor of cheese. These organisms are very resistant to the cooking temperature (≤38°C) used for Cheddar cheese but their growth in the cheese is limited to temperatures over 25°C. At temperatures below this, little growth and consequently little acid production occurs, but they can still supply enzymes that contribute to cheese ripening. In addition to affecting the flavor of cheese, they have a protective effect on acid production, as phage outbreaks for the mesophilic component of the culture will not affect acid production by the thermophilic component and vice versa. Like mesophilic cultures, thermophilic cultures also lyse during ripening, releasing their intracellular enzymes, whose activity improves the flavor of the cheese.

As noted above, adventitious NSLAB (i.e., those that are not added but are "accidental" contaminants from the cheese room or ripening chamber) grow to high numbers in ripened cheese. Because they are variable in species and number, they may be responsible for inconsistent cheese quality. Therefore, it is becoming increasingly common to add selected strains of NSLAB, chosen for their ability to improve the flavor of cheese, particularly Cheddar, with the starter. Selected strains of NSLAB, particularly *Lb. casei* and *Lb. rhamnosus*, may also be added to milk for Emmental cheese production to slow down the propionic acid fermentation. *Lb. rhamnosus* may also be added to milk in making Parmigiano Reggiano cheese to improve its flavor.

## Control of microbial growth in cheese

A block of hard cheese may be ripened for up to 2 years, raising the question as to why the cheese does not spoil or deteriorate during this time. The main reasons are the cooking temperature and the amount of time the curd is held at this temperature during manufacture, the pH of the cheese after manufacture, its salt and water contents, the presence of nitrate (in some varieties), and the relatively low ripening temperature.

The optimum temperature for the growth of bacteria is ~35°C for mesophiles and ~55°C for thermophiles. The maximum cooking temperature used in the production of semi-hard and hard cheeses range from 35°C (Gouda and Edam) to 54°C (Emmental and Grana), and the curd is also held for significant times around these temperatures. The lower temperature will promote the growth of mesophiles, including pathogens if present, while the higher one will inactivate many of them, especially as the pH of the curd is also decreasing due to the growth of the starter. Overall, the effect of cooking on contaminating mesophilic bacteria is probably fairly minimal.

The rapid production of lactic acid and the consequent reduction in pH and syneresis of curd, which, in turn, reduces the moisture content of the cheese, is an important factor. Most bacteria found in cheese prefer a neutral pH for optimal growth and, except for starters and NSLAB, will generally not grow below pH 5.0. The pH of cheese decreases rapidly during manufacture and immediately after manufacture ranges from 4.7 in the case of cottage cheese to 5.3 in the case of Cheddar and remains close to these values during ripening (cottage cheese is not ripened) and marketing. Surface-ripened cheeses are an exception to this, as the surface pH increases during ripening, often reaching values greater than 7.0. Organic acids like lactate, acetate, and propionate also act as inhibitors of microbial growth.

The use of salt to prevent microbial growth in food is probably as old as food production itself. The concentration depends on the composition of the food, but a useful rule of thumb is that 10% salt is sufficient to prevent deterioration of most foods. Most bacteria will not grow in this concentration of salt but some of those found on the surface of smear-ripened cheese, like corynebacteria and various species of yeast can; such microorganisms are termed halo- or salt-tolerant. The salt and water contents of cheese are intimately connected since the salt is dissolved in the water, which increases the effective salt concentration, e.g., typical values for salt and water contents in Cheddar cheese are 1.8% and 38%, respectively, giving an actual salt concentration of 4.73% (i.e., $1.8 \times 100/38$).

Microorganisms require water to grow, and many cheeses have relatively low moisture contents. However, it is the availability of the water, rather than the total amount present, that is important. Much of the protein in cheese is hydrated by binding water molecules and this "bound" water is not available for microbial growth. Water availability is expressed in terms of the water activity ($a_w$), which is defined as the ratio of the vapor pressure (the tendency of water in a material to transform into the vapor state) of the cheese divided by the vapor pressure of pure water at the same temperature; it ranges from 0 to 1. The presence of salt reduces the water activity, e.g., the water activities in the presence of 4% and 10% salt are 0.977 and 0.935, respectively. The minimum $a_w$ values for the growth of bacteria, yeast, and molds are 0.92, 0.83, and 0.75, respectively, but significant variation occurs. In addition, the $a_w$ of cheese decreases during ripening, due to evaporation of water from the cheese surface in cases where the cheese is not tightly packaged, and hydrolysis of proteins and fats by proteolytic and lipolytic enzymes during the ripening process. The water activities for cheese generally lie above 0.95, except for Parmigiano Reggiano and Sbrinz, which have typical values of 0.92 and 0.94, respectively. Traditionally, the latter two cheeses have long ripening times, sometimes up to 3 years.

Nitrate is added to milk for making some cheese types, e.g., Gouda and Edam, to prevent the growth of spoilage bacteria, particularly *Cl. tyrobutyricum*, the cause of late gas development in these cheeses. The actual inhibitor is nitrite, which is produced from the nitrate through the activity of the enzyme xanthine oxidoreductase in the milk. In the past, the development of early gas in cheese, caused by coliform bacteria, was fairly common, but this has become less important as a defect today, with improvements in raw milk quality and the use of active starters.

One further factor controlling the growth of microorganisms in cheese is the oxidation-reduction ($E_h$) potential, which is a measure of the ability of chemical/biochemical systems to oxidize (lose electrons) or reduce (gain electrons). It is measured in millivolts, using a platinum electrode. A positive value indicates an oxidized state and a negative value a reduced state. The $E_h$ of milk decreases during cheese-making from a positive value (+150 mv) to a negative value (−250 mv), due to the fermentation of lactose to lactic acid by the starters. The mesophilic starters *Ec. faecalis* and NSLAB are much better at reducing the $E_h$ than thermophilic cultures. At negative $E_h$ values, cheese is essentially an anaerobic (oxygen-free) system and obligate aerobes, like corynebacteria and *Micrococcus* spp., will not grow, except on the cheese surface where there is an adequate oxygen supply. In contrast, negative $E_h$ values will promote the growth of anaerobes in the cheese, like *Cl. tyrobutyricum*. The $E_h$ value will also affect the development of cheese flavor.

The temperatures used in cheese ripening ranging from 8°C to 12°C, are much lower than the optima for growth. The exception to this is Emmental, which is ripened at 22–24°C for some weeks early in the process to promote the growth of the propionic acid bacteria, after which the temperature is then reduced for the remainder of the ripening. This increase in ripening temperature could theoretically promote the growth of contaminating bacteria.

Individually, none of these will prevent microbial deterioration of cheese but all of them acting together, as so called "hurdles," do.

## Pathogens in cheese

Cheese is usually a very safe product. Between 1980 and 2013 there were only 53 food poisoning outbreaks due to cheese, while almost 250,000,000 tons of cheese were made worldwide. The most common organisms involved in these outbreaks were *Salmonella* (23 cases), almost all involving raw milk cheese, followed by *Listeria monocytogenes* (13 cases), enteropathogenic *E. coli*, particularly strain O157:H7 (7 cases), and *S. aureus* (4 cases). Of these outbreaks, *L. monocytogenes* is the most serious as it involved more than 100 fatalities and two major outbreaks in the 1980s, involving Mexican-type cheese (queso fresco) in California and Vacherin Mont d'Or, a smear-ripened cheese, in Switzerland. Common symptoms of food poisoning caused by these bacteria include diarrhea and abdominal pain. Even when these organisms are present in cheese, they will normally decrease during ripening if the cheese is well made. However, in smear-ripened cheeses, they increase during ripening, because of the high moisture content of many of these cheeses and the increase in pH that occurs on the surface during ripening. For Mexican-type cheeses, pH also plays a role in encouraging outbreaks, because it may not decrease very much during cheese manufacture.

*L. monocytogenes* is a particularly invidious microorganism because of its ability to grow at a low temperature (0°C), at low pH (4.4), and in relatively high concentrations of NaCl (10%–12%); some of these conditions often pertain in cheese during ripening and subsequent storage. The organism is widely distributed and is found in soil, silage, and water from various sources, including the surface water of canals, lakes, and rivers, and sewage. Some bacteriocins are particularly effective against listeria and efforts have been made to use them to reduce listerial growth on the cheese surface.[2]

One of the most important ways to control pathogens in cheese is to pasteurize the milk, which inactivates any pathogen that might be present (see Chapter 3). In addition to pasteurization, the other factors just mentioned also influence the safety of cheese. All of these so-called hurdles act together to give greater control of any pathogens and indeed of non-pathogens that might be present.

## Microbial spoilage of cheese

Eye formation, due to carbon dioxide production, is characteristic of some cheeses, e.g. Swiss types where propionic acid bacteria (PAB) ferment the lactate produced by the starter to propionate, acetate, and carbon dioxide, or in Dutch type cheeses where $Cit^+$ lactococci in the starter produce carbon dioxide and acetate from citrate present in the milk.

However, gas formation is also sometimes a cause of defects in cheese. Early gas production due to the growth of coliforms occurs in poorly made cheese and is characterized by the development of a large number of small eyes shortly after manufacture. This defect is not very common today due to improvements in the microbiological quality of milk and the use of active starters. Lactose retained in the cheese is the likely source of the carbon dioxide.

Late gas formation by anaerobic spore-formers, especially *Cl. tyrobutyricum* and *Cl. butyricum*, is much more common, especially in cheese like Gouda and Edam with relatively high moisture contents. The spores germinate, producing vegetative cells which then grow, fermenting lactate to butyric acid, hydrogen, and carbon dioxide. The source of these organisms is silage and, in some countries, e.g., Switzerland, feeding silage to cows is prohibited.

Some cheeses, especially aged Cheddar, can develop white spots late in ripening which are caused by the crystallization of amino acids and/or formation of calcium D lactate, from the L lactate. Some consumers do not consider this a defect but an indicator of maturity.

## **Raw milk cheeses**

Many cheeses are made from raw milk, including Parmigiano Reggiano, Roquefort, Comté, and Emmental (commonly referred to as Swiss cheese), and these varieties account for almost 10% of total cheese production in Switzerland, France, and Italy. Raw milk can be contaminated with pathogens, which can subsequently grow in it, but this does not happen very often, implying that cheese made from raw milk is relatively safe. Confounding issues, other than the use of raw milk, contributed to the growth and presence of pathogens in most of the outbreaks of food poisoning associated with raw milk cheese. The major problem was poor hygiene but, in some cases, the cows themselves shed the causative organism, *L. monocytogenes* or *Salmonella*, in their milk. In most countries today, milk is produced very hygienically, the starters used are active, and the cooking temperature in many cheeses made from raw milk is high, >54°C, e.g., Emmental, Grana cheese, Comté, and Gruyère; low pH, low moisture, and salting further retard the growth of the pathogens.

In the United States, the Food and Drug Administration (FDA) requires that cheese made from raw milk must be ripened for at least 60 days at >1.7°C before being sold, but this regulation has been questioned in many research studies showing that this time period is ineffective in ridding raw milk cheese of many pathogens. Application of the principles of Hazard Analysis of Critical Control Points (HACCP) (see Chapter 5) is important in making cheese from raw milk and indeed from pasteurized milk. Key control points in a HACCP plan for cheese might be the temperature to which the raw milk is cooled and the time it took to reach that temperature, the milk storage time (1 day at most), and the pH at some predetermined point after addition of the starter culture.

## Manufacture of some common cheese varieties

In the following sections, a brief description of the manufacture and principal characteristics of several common types of cheese is given.[3]

### Cheddar cheese

Cheddar cheese is thought to have originated in the village of Cheddar in Somerset, England, but there is no definitive proof of this; today, it is produced in many countries around the world. Traditionally, Cheddar cheese was produced in cloth-bound cylindrical shapes, but today it is produced in vacuum-packed rectangular blocks weighing about 25 kg. The cheese is ripened at 8–10°C for 3 to 6 months for a mild-flavored Cheddar or 1–2 years for a fully mature cheese. The milk is normally pasteurized at 72°C for 15 seconds, or 30 minutes at 63°C. On a small scale, the latter treatment can be conveniently done in a bulk tank, modified so that hot water can be circulated in the skin of the tank. The milk is then cooled to 32°C and rennet and a mesophilic starter (1–2% v/v) are added and gently stirred into the milk. The milk is allowed to settle for 30–40 minutes until it coagulates, and the coagulum is then cut. A simple way to determine if the coagulum is ready to be cut is to cut it to a depth of around 5 cm with a knife and then insert the blade under the cut and bring it up to the surface. If a clean break is observed, the coagulum is ready to be cut, vertically and horizontally, with "knives" consisting of blades, spaced 2–3 cm apart.

Cooking of the curd and whey is then begun, with more or less continuous stirring, and takes place over around 30 minutes until the temperature reaches 38–40°C. At the latter temperatures, acid production by the starter slows down. The objective of cutting the coagulum and the heating and stirring steps that follow it are to facilitate whey syneresis from the coagulum. Heating is then stopped, the curd is allowed to settle, part of the whey is drained off, and a central channel is made in the bed of curd to allow more whey to flow out. The curd granules fuse under gravity, and the mass of curd is then cut into rectangular slabs which are piled on top of each other as the whey continues to flow from the curd and out of the vat. This series of operations is known as cheddaring and is a characteristic feature of Cheddar cheese manufacture.

The principal stages of manufacture of Cheddar cheese are shown in Figure 6.5; the steps shown in the first three panels are in fact typical of the production of many cheese varieties.

The pH of the curd is monitored frequently; when it has decreased from its initial value of ~6.5 to 5.4, the curd is milled into 5 × 1 × 1 cm pieces and dry salted at a rate of 1.5–2%. The salt is mixed in thoroughly and the salted curd placed in molds, which have been previously lined with cloths; the cloths are then pulled tight before the curd is pressed for 16 hours. Cheddar cheese is ripened at 10–12°C for 6 months to 2 years, with the shorter ripening time generally giving a milder flavored cheese and the longer ripening time giving a more mature and strongly flavored cheese.

Over the past 50 years significant efforts have been made to remove the labor-intensive steps of making Cheddar through mechanization, including improvement in vat design, development of belts and towers in which cheddaring occurs automatically, development of automatic salting and guillotines to cut the cheese to the required weights, automatic bagging and packaging systems, and automatic transfer of the packaged cheese to the ripening rooms. A block of freshly made and packaged Cheddar cheese and a slice of a mature Cheddar cheese, freshly sliced, are shown in Figure 6.6.

Figure 6.5 Various stages of Cheddar cheese-making, showing (clockwise, from top left) (a) cutting the milk gel, (b) stirring the mixture of curds and whey, (c) separated curds, and (d) cheddared curd blocks (images thanks to David Waldron).

**Figure 6.6** (a) A vacuum-packed block of freshly made Cheddar cheese, bland in flavor and rubbery in texture and (b) a portion of a mature farmhouse-made Cheddar cheese with a natural rind (image thanks to David Waldron).

## Gouda and Edam cheese

These mild, slightly sweet-tasting cheeses originated in the Netherlands and, although different in shape and surface color—Edam is spherical, covered with a red wax, and weighs 1–4 kg, while Gouda is cylindrical, covered with a yellow wax, and weighs 4 to 6 kg—are made in a similar way (Figure 6.7). Normally, the milk is standardized to a fat content of 3.5%. After pasteurization at 73°C for 15 seconds, the milk may be bactofuged, as described in Chapter 5, to remove clostridial spores, which can germinate and grow in the cheese during ripening, causing late gas formation. In the past, sodium or potassium nitrate was added to prevent the growth of clostridia, but this is uncommon today because of the valuable nature of whey as a source of ingredients for other products, especially infant formulae.

After pasteurization, the milk is cooled to 30°C and a diacilactis leuconostoc (DL) starter (see Chapter 4), calcium chloride (to improve rennet coagulation) and rennet are added. Cutting of the coagulum into pieces of 8–15 mm begins 20–25 minutes after adding the rennet; this is done very gently to reduce losses of fat and protein "fines" in the whey. The whey/curd mixture is heated to ~35°C and, when the pH has decreased to 6.45, stirring is stopped and the curds allowed to settle. A percentage (around 25%) of the whey is pumped off and replaced by

**Figure 6.7** A slice of mature Gouda cheese, showing the distinctive wax coating, a single small eye, and some crystals, possibly of calcium phosphate or the amino acid tyrosine, near the rind (image thanks to David Waldron).

hot water (at ~55°C); for this reason, these varieties are sometimes referred to as washed-curd cheeses. The curds form a continuous mass, which is traditionally cut into square blocks, each of which is then placed in a round mold; each square block of curd spreads to fill the mold, and the molds are then pressed.

Acidification continues during pressing and the pH decreases from 5.9 at the beginning to 5.4 within 1.5 hours. During pressing, the cheese loses considerable moisture, and a rind forms; the cheeses are turned once during this time, and all the remaining lactose is converted to lactic acid. The cheese is then brined in 17–18% NaCl for 5 days. The average salt content of a young Gouda ranges from 4–5 g/100 g of moisture, but it can take up to 2 months for the salt to become evenly distributed throughout the cheese. A wax coating, to which an antibiotic, natamycin, may be added to inhibit the growth of yeast and molds, is applied to the surface of the cheese before ripening. The cheese is ripened at 12–15°C and 85–88% relative humidity (RH), for up to 12 months, for a mature Gouda.

## Emmental cheese

The classic Swiss cheese, Emmental; originated in the Emmen valley in the Canton of Bern in Switzerland. These cheeses have a characteristic nutty flavor and are traditionally large, cylindrical cheeses, weighing 60–130 kg and characterized by the development of large holes, called "eyes," in the cheese during ripening. These eyes are due to the growth of, and production of carbon dioxide by, PAB, especially *Propioibacterium freundenreichii*, small amounts of which are added to the milk with the lactic starter. Small amounts of lactobacilli, particularly *Lb. casei* and *Lb. rhamnosus*, are also included in the starter, the function of which is to slow down the growth of the PAB in the cheese. The eyes are generally round, 1–4 cm diameter, with about 1,000–2,000 present in each cheese; they usually form where tiny particles of dust or hay are entrapped in the curd. The sequence of operations in the manufacture of Emmental cheese is geared to optimizing conditions for the growth of the PAB.

This type of cheese is made from raw milk from cows that are not fed silage; this reduces contamination by clostridia, especially *Cl. tyrobutyricum*, which may cause late gas formation, due to the production of hydrogen, carbon dioxide, and butyric acid from lactate in the ripening cheese. Accumulation of butyric acid can also give the cheese a rancid flavor. A thermophilic starter comprising mixed cultures of lactobacilli (*Lb. helveticus* and *Lb. delbrueckii* subsp. *lactis*) and *Sc. thermophilus* is used, which converts the lactose to lactic acid. Galactose accumulates early during ripening, because *Sc. thermophilus* is unable to metabolize it and excretes it in proportion to the amount of lactose transported into the cell. However, it is eventually metabolized by *Lb. helveticus*.

Traditionally, copper vats are used. A portion of the copper leeches into the cheese and stimulates growth of the PAB; however, too high a concentration can inhibit their growth. In addition, 12–18% of water may be added to the milk in the vat to reduce the lactose level, which leads to a relatively high pH (5.2–5.3) in the curd and promotes the growth of PAB.

After coagulation, the coagulum is cut into small pieces using a special type of knife called a Swiss harp and the whey-curd mixture is heated to 52–54°C, during which the coagulant is inactivated, and curd particles shrink to the size of rice grains. Part of the whey is run off, and the curd is placed in large molds and pressed for several hours. Because of the large mass of curd in the mold, the temperature remains above 50°C for several hours during pressing and, as a result, much of the lactic fermentation occurs while the cheese is being pressed. The decreasing pH and higher temperatures act as further deterrents of growth of any pathogens that may have been present in the raw milk. Once a good rind has been formed, the cheese is placed, for up to 2 days for large wheels, in saturated brine (23% NaCl). PAB are sensitive to salt, so the brining time for Emmental cheese is less than for other cheeses and the content of salt in the cheese is lower, at 3–5 g/kg of cheese, than in other varieties.

A few days after manufacture, the cheese is transferred to a "hot room" at 20–24°C for 40–60 days to allow the PAB to grow, during which time they ferment lactate to propionate, acetate, and $CO_2$. The propionate gives the cheese its characteristic nutty flavor, while the $CO_2$ is responsible for eye formation; visible formation of eyes takes several weeks (sometimes up to 4). In a cheese weighing 80 kg, the total amount of $CO_2$ produced is about 120 liters when the cheese is ready to eat. Half of this is dissolved in the cheese, 20 liters are found in the eyes and the remaining 40 liters diffuse out of the cheese during ripening. Once the propionic acid fermentation is complete, the temperature is reduced to 10–13°C and the RH to 70%–80% to continue ripening. The final appearance of a large piece of ripened Emmental cheese is shown in Figure 6.8.

## Grana Cheeses

Parmigiano Reggiano and Grana Padano are the two main Italian "grana" cheeses, so-called because of the granular structure of the ripened cheese. They have a diameter of 33–45 cm and a weight of 35–37 kg, a low moisture content (~30%) and a distinct sweet flavor due to their long, slow ripening. Silage feeding to cows is not permitted if their milk is to be used for Parmigiano Reggiano but is allowed for the production of Grama Padano. These cheeses are only produced on individual farms in designated areas of Northern Italy, but the cheeses are sent to a centralized facility for ripening.

(a)

(b)

**Figure 6.8** (a) Half a wheel (about 90 kg in weight), a wedge, and a block of Emmental cheese (Consortium Emmentaler AOP, Berne, Switzerland) and (b) slices of a large wheel of Emmental cheese; both pictures show the distinctive large eyes formed by the production of carbon dioxide from lactate by propionic acid bacteria during the warm room ripening phase (image thanks to David Waldron).

For Parmigiano Reggiano, the previous evening's raw milk is partially skimmed by holding it overnight at 20°C in shallow tanks before removing portion of the cream. During this process, some acidification of the milk occurs. The partially skimmed milk is mixed with the next morning's raw milk in a ratio of 1:1 so that the overall fat content is around 2.5%. This step is skipped in the production of Grana Padano. In both cheeses a natural whey culture (3%), produced from the previous day's whey, which has been held for 24 hours at temperatures between 35 and 50°C, is used as a starter; the whey culture contains a complex mixture of LAB, dominated by lactobacilli. Calf rennet is used as coagulant and the maximum cooking temperature ranges from 53–55°C. The time from addition of rennet to the end of cooking is short, about 22 minutes, and the coagulum is cut with a special cutter called a spino. The vats used for Parmigiano Reggiano production are shaped like an inverted bell and have a capacity of 100–120 liters. Two cheeses are produced from each vat, which, once drained through cloths, are placed in individual molds and pressed. They are then left in saturated brine for 20–24 days and are ripened for 18–24 months (Parmigiano Reggiano) or 12–16 months (Grana Padano) at ~18°C and a RH of ~85%.

The appearance of ripening and cut Parmigiano Reggiano cheese are illustrated in Figure 6.9.

## Pasta filata cheese

Pasta filata cheese is an important product in Southern Europe; in Italian, the name means "stretched curd," referring to the plasticization and stretching steps used in its manufacture. Examples include mozzarella, kashkaval, provolone, and caciocavallo. Mozzarella, especially low moisture mozzarella (LMM), is by far the most important variety because of its use as a topping for pizza. Although traditional mozzarella cheese is made from raw, water buffalo milk (Mozzarella de Bufalo) in southern Italy, most LMM is made from pasteurized cows' milk. LMM is often called pizza cheese since its key functional properties relate to its ability to flow, stretch, and brown when baked on a pizza (as seen in Figure 6.10).

The milk used is standardized to a casein:fat ratio of 1.2:1, normally by adding skim milk powder to the milk or, less frequently, by removal of some of the cream. The milk is then pasteurized at 72°C for 15 seconds. The basic steps in the manufacture of LMM are very similar to those for Cheddar cheese up to the milling step, except that a thermophilic starter (Chapter 4) is used rather than a mesophilic one. Rennet is used to coagulate the milk and, after the coagulum is cut, the curds and whey are cooked at 41°C (when a thermophilic starter is used), after which part of the whey is drained off. The remaining curds and whey are pumped onto an enclosed conveyor belt system, where further whey

(a)

(b)

**Figure 6.9** (a) Ripening Parmigiano-Reggiano cheese showing the typical bulge and manufacturing codes (thanks to Consorzio del Formaggio Parmigiano Reggiano) and (b) a slice of this cheese grated for service (image thanks to David Waldron). Crystals are again clearly visible due to the low moisture content, especially at the cheese edge.

**Figure 6.10** Mozzarella cheese should melt, flow, and stretch into fine strands when cooked on a pizza.

drainage and curd matting proceed until the pH reaches 5.1–5.3, if the curd is to be milled and stretched immediately; if not, the curd may be milled at a slightly higher pH and then dry salted. Acid production is more rapid in LMM than in Cheddar cheese, which reduces the make time to 2.5 hours compared to about 5.5 hours for Cheddar.

The salted or unsalted curd is then stretched (plasticized) mechanically in hot water or hot brine. Stretching is a two-stage process. During the first stage, the milled curd is placed in a tank of hot water, where its temperature increases as it falls to the bottom of the tank. At 50–55°C, the curd is converted to a plastic and workable consistency. In the second stage, the plasticized curd is kneaded and stretched into a fibrous ribbon of hot plastic curd, which is then forced under pressure into chilled molds. This cools the cheese so that it retains its shape when removed from the mold.

The cheese is not ripened, as flavor development is not an important consideration for its use on pizza, which is usually strongly flavored itself. However, it is refrigerated after plasticization and typically achieves its functional properties within a month, mainly due to the absorption of water by the curd, which gives it a particular fibrous structure that is linked to its desirable properties of flowing and stretching.

## Blue cheeses

Blue cheeses are characterized by the development of blue veins throughout the cheese mass during ripening, due to the growth of the mold, *Penicillium roqueforti*, the conidia (spores) of which are added to the milk with the starter. They are made in many countries throughout the world and come in various sizes. The most famous are Roquefort (France), Gorgonzola (Italy), Stilton (United Kingdom), Danablu (Denmark), and Blue cheese (United States), most of which have Protected Designation of Origin (PDO) status. Roquefort is produced from sheep milk and, to be called Roquefort, must be ripened in the caves at Roquefort-sur-Soulzon. France. Stilton can only be produced in the adjacent counties of Derbyshire, Leicestershire, and Nottinghamshire in the United Kingdom and Gorgonzola in some provinces of northern Italy. The appearance of a mature blue cheese is shown in Figure 6.11.

Blue cheeses are made from cow, sheep, or goat milk, or a mixture of two or more of these milks The starters used can be either mesophilic or thermophilic.

**Figure 6.11** Mature blue cheese, showing the extensive growth of *Penicillium roqueforti* and the crumbly texture of the ripened cheese (image thanks to David Waldron).

The milk can be raw (Roquefort) or pasteurized (Gorgonzola and Stilton). Mold conidia, rennet, and starter are added at 32°C, though sometimes the starter is omitted, if the cheesemaker relies on adventitious LAB present in the milk to produce the necessary lactic acid. After coagulation and cutting of the coagulum, the curds are placed in the molds without a heating step, and the curd is allowed to drain freely, without pressing, for 2–3 days, with frequent inversion of the molds. After drainage, the pH can range from 4.6–5.2 and, depending on the variety, the cheese is ripened at 8–15°C and 85–95% RH. Salting can either be by immersing in brine or applying dry salt to the cheese surface. Both methods create a salt gradient from the cheese surface to its core. The concentration of salt in blue cheese ranges from 1.6–5%.

Sometimes, harder blue cheeses are pierced with stainless steel skewers during ripening to allow oxygen to penetrate the interior to promote the growth of *P. roqueforti*. Several different species of yeast, such as *Yarrowia lipolytica*, *Debaryomyces hansenii*, *Kluyveromyces lactis*, and *Geotrichum candidum*, have been reported on the surface of different varieties of blue cheese. They are thought to originate in the raw milk if this is used (as pasteurization kills yeast), or in the brine or environment in the case of blue cheese made from pasteurized milk. Considerable lipolysis occurs in blue cheese due to the lipolytic activity of *P. roqueforti* and, in turn, the free fatty acids are converted to methyl ketones which dominate the flavor of these cheeses

## Mold surface-ripened cheeses

The most important mold surface-ripened cheeses are Camembert and Brie, both of which can be made from either raw or pasteurized milk. Camembert de Normandie and Brie de Meaux are raw milk cheeses, characterized by the development of a soft, white to grey, velvety appearance on the surface—the so-called bloom caused by the growth of *Penicillium camemberti*. Immediately beneath the bloom is a dark orange or brown layer, about 0.5 mm thick, composed of yeast and bacteria. Camembert is a flat, cylindrical cheese, at least 10 cm in diameter and 2.5 cm thick and at least 250 grams in weight. Brie is a similar looking cheese but with a much larger diameter (27–30 cm).

Regardless of whether they are made from raw or pasteurized milk, mesophilic DL starters are used in the manufacture of both cheeses. When the milk is pasteurized, *Pen. camemberti* is added with the starter, unless raw milk is used, as the *Penicillium* spores originate in the milk itself. Yeast (*D. hansenii* and *Geotrichum candidium*) may also be added. The milk is coagulated by rennet at ~32°C in 30–45 minutes, after which the coagulum is cut into cubes. The coagulum is not cooked but is ladled into the molds (5 ladles/per mold) and the

**Figure 6.12** Camembert cheese, showing the mold growth on the surface and the soft interior (image thanks to David Waldron).

whey drains through perforations on the sides of the molds. Initially, the temperature is kept at 26–28°C, but is then gradually reduced to ~20°C. At the end of draining, the pH of the curd is ~4.6, the isoelectric pH of casein. The cheese is dry-salted and ripened at 11–13°C and 90% RH for 21 days; a complete cheese and slice of ripe Camembert are shown in Figure 6.12.

The composition and evolution of the surface flora of Camembert cheese during ripening is complex, particularly when raw milk is used, and includes, as well as the starter bacteria, mold (*Pen. camemberti*), other bacteria (*Brevibacterium linens* and corynebacteria), and yeast (*Geotrichium candidium*, *D. hansenii*, and *Kluyveromyces lactis*). In raw milk cheese, the physicochemical parameters, e.g., salt, $a_w$ and pH, select the microorganisms that grow, whereas in pasteurized milk, cheese most of the flora required, including the mold, yeasts, and the bacteria, are added to the milk with the starter. The yeasts and molds grow during the first 5–10 days of ripening, metabolizing the lactate produced by the starter and increasing the pH on the cheese surface.

## Bacterial surface-ripened cheeses

Bacterial surface-ripened cheeses are also called red smear or washed-rind cheeses and are characterized by the development of an ill-defined mixture of yeast and bacteria on the surface during ripening, which gives the cheese a red color. These cheeses vary in size and shape and appearance. They can be hard, e.g., Gruyère and Comté, semi-hard, e.g., Tilsit and Limburger or soft, e.g., Münster, Livarot, and Reblochon. Most washed-rind cheeses are brine-salted. However, in the manufacture of Comté, salt and smear microorganisms are rubbed onto its surface several times a week during the first 3 weeks of ripening.

Hard, surface-ripened cheeses are made with thermophilic starter cultures; semi-hard and soft varieties are made with mesophilic ones. Cheeses made with thermophilic cultures are cooked to around 54°C. Cooking of washed-rind cheeses made with a mesophilic culture is limited (to ~35°C), so consequently these have a relatively high moisture content. After light pressing, sometimes overnight, the cheeses, whether made with thermophilic or mesophilic starters, are brined (in saturated brine, at pH 5.2, containing calcium) for 4–18 hours depending on their size; smaller cheeses are brined for a shorter time than larger ones. Sometimes, the only pressing comes from the weight of the curd itself.

The cheeses are then drained for a few hours, after which the surface is repeatedly washed with an aqueous suspension of yeast (*G. candidum* and *D. hansenii*) and bacteria (*B. linens*). However, this practice has been questioned since the strains added show poor recovery from the cheese during ripening and are quickly overtaken by a "house" microflora of contaminating microorganisms from the cheese-making environment. The "old-young" method of smearing may also be used: a smear is washed off the surface of a ripened (old) cheese and used to inoculate the surface of the fresh cheese. Smearing is usually done 2 or 3 times at 2–4-day intervals from the beginning of ripening. Between intervals, microcolonies that grow on the cheese surface are spread evenly by the washings, resulting in development of a more uniform smear. The cheese is ripened for 2–3 weeks at 10–15°C at a RH greater than 90% to allow the surface microflora to develop and produce the red or orange color. Both soft and semi-soft cheeses are then wrapped or transferred to another ripening room at a lower temperature for further maturation; some surface-ripened cheeses can get very soft during ripening, as illustrated in Figure 6.13.

The surface of washed-rind cheese has a relatively high salt content and a low pH ~5.2, and therefore the microorganisms which grow on it are salt- and pH-tolerant. Traditionally, it is thought that yeasts grow first, raising the pH through the utilization of lactate, which in turn, allows the bacteria to grow; however, some surface bacteria can also metabolize lactate and increase the pH of the surface.

The smear of these cheeses is diverse and complex. The sources of the yeast and bacteria in smear cheeses include the brine, the arms and hands of the personnel involved in cheese-making, and the wooden utensils, vats, and shelves, used in cheese manufacture and ripening. During ripening, cheese is in contact

(a)

(b)

**Figure 6.13** Two stages of maturity of surface-ripened cheese, (a) young and (b) mature, showing the progressive softening of the interior during ripening (images thanks to David Waldron).

with wooden shelves for a considerable time, raising the possibility that they may be a source of pathogens. The old-young method can also result in contamination of the young cheese by pathogenic bacteria, particularly listeria.

## Feta-type cheese

Feta is a popular cheese that originated in Greece but is now produced worldwide. It is usually made from sheep milk or a mixture of sheep and goat milk and is ripened in brine. In large scale production the milk is normally pasteurized. Traditionally, yoghurt cultures or a 24 hour-old yoghurt was used as the starter but nowadays the *Sc. thermophilus* component of the yoghurt culture has been replaced by *Lc. lactis*. Calf rennet is used to coagulate the milk in about 50 minutes at 32°C. The coagulum is cut into 2–3 cm cubes and left for 10 minutes for initial whey expulsion before the curds are ladled into perforated molds where further whey drainage occurs. After 2–3 hours at 14–16°C, the molds are inverted and left for a further 2–3 hours to complete whey drainage.

The curd is then removed from the mold and cut into several pieces of equal size. These are placed on a table covered in coarse salt, and their upper surfaces are also covered with salt. Every 12 hours, the pieces are inverted until the salt level in the curds reaches ~3.5%. The pieces of salted curd are then left on the table for a few more days to allow an adventitious slime of bacteria, yeast, and molds to develop, which is then washed off with brine. The salted pieces of curd are then tightly packed in tin-plated cans or plastic packaging, with any excess space filled with 6–8% brine, and ripened at 16–18°C for 2–12 months. The final pH attained in the cheese is 4.5, and final ripe cheese looks as shown in Figure 6.14.

## Cottage cheese

Cottage cheese is a soft, unripened cheese, produced in many countries from skim milk or reconstituted skim milk (Figure 6.15). The main principle behind making cottage cheese is to coagulate the casein at, or near, its isoelectric pH of 4.6. Regardless of the type of milk used, it is normally pasteurized at 72°C for 15 seconds; while higher temperatures (75°C) may be used, this can result in a soft coagulum and poor curd syneresis. DL starters are used, as the diacetyl produced by such cultures is an important flavor component of this cheese. A small amount of rennet is also added, though significantly lower than that used for the types of cheese discussed above. After cutting the coagulum and whey drainage, incubation continues until the pH reaches 4.6.

**Figure 6.14** Feta cheese, a cheese ripened and stored in brine, some of which is clearly visible on the board (image thanks to David Waldron).

**Figure 6.15** Cottage cheese, a very high moisture cheese, essentially comprising acid-coagulated casein (image thanks to David Waldron).

The pH may also be reduced through direct acidification with GDL, or a mixture of lactic or phosphoric acids and GDL. In the latter case, the acid is added to the milk at 2–12°C to reduce the pH to 5.2, without coagulating it, after which GDL is added and the temperature increased. Skim milk at pH 5.2 and 27°C requires between 4.8 and 5.4 g GDL to reduce the pH to 4.6 within 1 hour.

The pH at which the coagulum is cut is the most critical step in the manufacture of cottage cheese. The optimum is pH 4.8, after which the curds and whey are allowed to synerese for about 15 minutes, during which the pH decreases to 4.6. Then the curd is cooked at 52–60°C, the whey is drained off, and the curd washed 2–3 times with potable water both to remove excess lactose and lactic acid and to cool it. During the washing step, the hot water should be drained from the vat jacket so that the temperature of the curd does not increase further.

The final step in cottage cheese manufacture is adding a cream dressing. A typical formula for the dressing is 10.4% fat, 7.6% solids-not-fat, 2% whey, 2.5% NaCl, and 0.35% starch or guar gum to stabilize the product. These are blended together, pasteurized at 74°C for 30 minutes, homogenized, and cooled to 4°C. Alternatively, the temperature may be held at 22°C and the mixture inoculated with a DL culture and incubated for 6 hours. Regardless of the latter step, the cream is mixed with the curds in a ratio of around 40:60.

## Queso fresco

Queso fresco (fresh cheese in Spanish) or queso blanco (white cheese in Spanish) are popular in Central and South America and in areas of the United States where there are large Spanish-speaking populations. It differs from most other cheeses in that it is generally made from raw milk and no starters are used; it has a high moisture content and is not ripened. Sometimes, no rennet is used, and instead coagulation is by direct addition of lactic or acetic acids or even lemon juice in artisanal production. A survey of nine commercial brands of queso fresco made in the United States showed significant compositional variation in pH (6.63–6.86), moisture (44–64.5%), NaCl (1.53–2.01%), and levels of lactic acid bacteria.[4] The high pH and moisture levels and relatively low levels of salt for that level of moisture imply that such cheeses are prone to the growth of pathogens, and indeed this has occurred. This cheese has been involved in at least three major outbreaks of listeriosis in the United States, one in California in 1985 involving 140 cases, including 48 deaths, one in North Carolina in 2000 involving 12 cases, most of whom consumed queso fresco made on one dairy farm, and one in Texas in 2003. There was no indication of how old the cheeses were, except that they were within their sell-by date.

## Paneer

Paneer is an Indian milk product produced by acid coagulation of buffalo milk, which is unusual in being acidified with an acid, usually citric, rather than with a starter culture.[5] Buffalo milk containing 6% fat is heated to 82°C in a cheese vat for 5 minutes, cooled to 70°C, and coagulated with 1% citric acid, which is added slowly to the milk with continuous stirring until the curd and clear whey separate. The mixture is allowed to settle for 10 minutes, and the whey is drained through a muslin cloth at 63°C. The curd is then collected and poured into a rectangular-framed mold (35 × 28 × 10 cm), lined with a clean muslin cloth. The mold is open at both top and bottom and placed on a wooden plank. Once filled with the curd, it is covered with another piece of wood, on top of which a 45 kg weight is placed for about 15–20 minutes. The pressed block of curd is then removed from the frame, cut into 10–16 cm pieces, and immersed in pasteurized chilled water (4–6°C) for 2–3 hours. The chilled pieces of paneer are removed from the water and placed on a wooden plank for 10–15 minutes to drain occluded water. Afterward, they are wrapped in parchment paper and stored at refrigeration temperature.

## Whey cheeses

Cheese may also be made from whey, e.g., ricotta from sheep milk whey in Italy and mysost from cow milk whey in Norway. In making ricotta, the whey or a whey-milk mixture is acidified to pH 5.6–6.0 with a starter or directly with acid (typically acetic or citric). The mixture is then heated to 80°C for a few minutes to flocculate (by denaturation) the whey proteins, which are then scooped into molds to drain off the protein-depleted whey. Salt is sometimes added to the whey before heating to enhance flocculation.

Mysost is essentially highly concentrated evaporated whey to which milk fat, cream, or goat milk, in the traditional version, is added. Crystallization of lactose during storage is a major problem in its manufacture, because of its relatively low solubility. After evaporation, the mixture is cooled rapidly in a special kettle, called a "gryta" in Norwegian, from 93°C, the final temperature of evaporation, to 65°C to prevent lactose crystallization. An important feature of mysost production is that there is no whey by-product, in contrast to ricotta, where the protein-depleted whey requires disposal.

## PDO cheeses

The localized production of certain varieties of cheese in Europe is now protected and encouraged through the Protected Designation of Origin (PDO) or

Protected Geographical Indication (PGI) programs. These areas are delineated by European law that came into effect in 1992 in an effort to protect regional food production. The regulations define the region and manufacturing technology for certain cheese varieties and indeed other foods. The first cheese to get PDO status was Roquefort, and since then only cheese made from raw ovine milk and ripened in caves in the Roquefort-sur-Soulzon region of southern France can bear that name. PDO cheeses are numerous, especially in Italy (e.g., Parmigiano Reggiano, Grana Padano, Mozzarella di Bufala, and Gorgonzola), France (e.g., Camembert di Normandie, Brie de Meaux, Roquefort, Beaufort, and Cantal) and Spain (e.g., Queso de la Serena, Idiazabel, Cabrales, and Mahon) and many are made from raw milk. Eleven cheeses in the United Kingdom have PDO status, including Stilton, Buxton Blue, and Swaledale; Ireland has just one, Regato. A similar system, called Appelation d'Origine Contrôlée, operates in France for French artisanal foods.

## Processed cheese products

Attempts to increase the shelf-life of cheese in the early 20th century were inspired by the possibility of increasing trade via the production of more stable cheese products and the existence of heated cheese dishes such as cheese fondue. Early attempts to produce a stable heated cheese product were unsuccessful, because the products were unstable due to oiling off and moisture exudation during cooling and storage. In 1911, two Swiss researchers, Walter Gerber and Fritz Stettler of the Gerber Company in Thun, produced a stable heat-treated Emmental cheese product by adding sodium citrate (referred to as "melting salt") to the grated and minced cheese before processing. James Lewis purchased a small cheese factory in Stockton, Illinois, in 1914 and commenced the production of processed Cheddar cheese in 1915; this was the start of the Kraft Cheese Company. Today, global production of processed cheese products is estimated to be about 2 million tons per annum, representing about 10% of natural cheese production.

Processed cheese originated as a way of utilizing natural cheese that was not of ideal quality or flavor or did not meet the required composition, but through processing could be converted into a lower-cost but still valuable alternative product. Today, in some countries, such as the United States, sales of processed cheese (particularly for applications such as cheeseburgers) exceed those of natural cheese.

The principle of production of processed cheese involves a radical redesign of the structure of cheese, through a process that generates a smooth and flexible structure, that can, during manufacture, be formed into a range of shapes such as blocks, slices, wheels, and triangles, and which has a long shelf-life without noticeable deterioration in quality (properties that are highly valued for certain applications). To understand this change in structure, consider what happens when Cheddar cheese is melted. This cheese contains three main components, protein, fat, and water, which do not mix naturally. When cheese is heated to the point of melting, these components separate into three phases: insoluble protein at the bottom, fat, which rises to the surface, and a murky watery phase in between.

Many food products contain oil and water in the form of an emulsion; in such cases, oil and water are mixed in the presence of a third component, an emulsifier, which acts as a bridge between the hydrophobic oil and the hydrophilic water, with a structure containing elements with both properties. In many food systems, protein acts as an emulsifier, as many proteins have the right blend of hydrophobic and hydrophilic regions (i.e., they are amphiphilic). The protein in cheese is casein, which has been destabilized by the action of rennet and cross-linked into a complex network by calcium ions; this is known as paracasein or calcium paracaseinate. This structure is too entangled to unfold readily at the surface of the separated fat in melted cheese. The problem is solved by chelating the calcium ions using sodium or potassium phosphate or sodium or potassium citrate, which have a high affinity for calcium ions; when added to melted cheese, they bind these ions, drawing them out of the casein structure and changing the protein present from calcium paracaseinate to sodium or potassium paracaseinate. Critically, whereas calcium paracaseinate is a very poor emulsifier, these new forms are very good emulsifiers and if, after adding these salts, vigorous mixing is applied to disperse the fat, the sodium or potassium paracaseinate coats the resulting droplets, providing a physical barrier to coalescence and producing an emulsion.

The principle of production of processed cheese involves taking cheese, adding extra water, melting the mixture in the presence of the emulsifying salts, mixing vigorously (to disperse the fat into droplets and cover them in the newly created protein emulsifier), and cooling the resulting emulsion, whereupon the fat present, and hence the whole product, solidifies. The fact that this final stage of production involves the conversion of a fluid to a solid gives important flexibility to making processed cheese products, as the molten product will take the shape of the package into which it is poured, such as a block or a triangle. In addition, the process of making processed cheese involves sufficient heat to inactivate most of the enzymes and microorganisms present, rendering the final product relatively sterile and unchanging (unlike the dynamic system in the parent cheese) and giving it a long shelf-life.

The production of processed cheese starts with selecting cheeses and preparing them for rapid melting and mixing by grinding them to a fine consistency using rollers and mincing extruders. A key question is exactly what cheeses should be used. As discussed above, cheese undergoes profound changes during ripening after manufacture, and the physicochemical properties of protein, the critical ingredient for making the new emulsion, change as the cheese ages. Young cheese contains a lot of intact casein, which gives a firm body and consistency to the processed cheese, but because it has not been broken down the cheese will be bland and lacking in flavor. On the other hand, older cheese, which has been ripened for several months, will have developed more flavor (the older, the stronger), but contains much less intact structure-forming casein; it will also be more expensive, owing to the cost of ripening.

The selection will depend to a large extent on the end use of the cheese. In some applications, a firm texture (young cheese) is more important than flavor, while in others a strong cheese flavor is very desirable. For example, for processed cheese intended for incorporation into other strongly flavored food systems, e.g., a slice on a cheese burger, the ability to slice cleanly and retain its shape in a hot, combined product is more important than a delicate flavor, as the cheese is part of a complex flavor profile, including meat, tomato sauce, onion, and several other strongly flavored components. In contrast, a spreadable processed cheese intended for use on crackers is valued by consumers more for its cheesy flavor, and strong structure is less important.

Thus, every processed cheese product, with a defined end-use and target application, will require a different type of starting material, and in all cases, this means selecting a blend of young, medium-mature, and extra-mature cheese, with the relative proportions of each depending on the considerations mentioned above.

When selected, the various cheese materials are grated and refined to a fine consistency and typically placed in a large bowl, called a kettle. To this is added the next critical ingredient, the emulsifying salt, which may be selected from a menu of complex inorganic chemicals, including polyphosphates, cyclic phosphates, citrates, and/or hexametaphosphates. Suppliers of emulsifying salts will provide blends that are recommended for the specific type of processed cheese being made, with the right blend of properties, such as the ability to not only convert the insoluble casein in cheese into the emulsifier, sodium caseinate, but also their influence on the product pH, solubility, in some cases antimicrobial activity, and other attributes.

The next ingredient is water, some of which is required to dissolve and allow easy incorporation of the emulsifying salts, and much of which is added because one of the traditional advantages of processed cheese is reduced cost compared to natural cheese.

Other ingredients that might be added include other sources of milk fat and protein (e.g., butter, anhydrous milk fat, and skim milk powder), and also color such as annatto (if making a strongly colored product from a pale cheese blend) and flavor components not traditionally found in cheese, such as ham and onion. When all the desired ingredients have been added in the right proportions, the lid on the kettle is sealed, and the mixing unit, which rotates at high speeds to shear and cut through the mass, is turned on, helping to dissolve the cheese and then dispersing the released and melted fat into droplets that will become coated in protein.

Heating is required to melt the fat and reduce viscosity, which can be achieved through the introduction of steam into the kettle's hollow wall, via electrically heated walls, or by steam piped directly into the molten cheese mass. The direct heating method (analogous to direct UHT treatment, see Chapter 5) results in very rapid heat transfer without the risk of any cheese burning on or sticking to the hot walls of the kettle. The heat transfer occurs through the steam condensing and releasing its latent heat of condensation; this adds water to the product, which must be taken into account in calculating how much extra water to add.

The mixture continues to be heated for a period of time after the final heating temperature has been reached. The exact conditions used depend not only on the minimum temperature required to inactivate most of the microorganisms and enzymes present (typically >70°C for 5–10 minutes), but also on the desired final product properties. One consideration in this regard is the required consistency of the cooled processed cheese and the impact this has on the quality of the emulsion produced. Any emulsion will destabilize if the droplets of oil or fat therein separate by rising vertically through the product, due to gravity and density differences (see the discussion of Stoke's Law in Chapter 2). If the product is solid (or close to it, as in a block of processed cheese), then the viscosity of the environment surrounding the droplets is so high that the droplets are basically trapped in place by their rigid surroundings. In the case of a much more liquid spread, however, the risk of separation is much greater, as the whole system is more mobile, and the process needs to "work harder" to distribute the fat in very small droplets that, due to Stoke's Law, are slow to separate. Thus, spreads are typically heated to a higher temperature (up to 95°C) and mixed more vigorously and longer than a mix intended to form a semisolid block of processed cheese.

When the required heating period has elapsed, the processed cheese mix will resemble a very viscous, hot paste. The next step is to cool it and allow it to solidity into its intended final form. Herein lies one of the key advantages of processed cheese in that, like water when it freezes, it will occupy the shape of whatever container it is placed in when it solidifies. So, to produce cheese in its final shape requires only that it be pourable, into a mold or package of the desired shape, while still hot, within which it "sets." The slices commonly used in convenience food products like burgers can be produced in a number of ways, including pouring the molten mix into slice-shaped sleeves, producing a block and then slicing it, or pouring it onto wheels or belts on which it is sliced.

During the cooling process, one of the key changes that occurs, and which gives the final product solidity at refrigeration temperatures, is solidification (through crystallization) of a high proportion of the milk fat present; as discussed in Chapter 2, this can give different structures and textures depending on the crystal forms, shapes, and sizes formed, which is, in turn, determined by the cooling rate, and so the rate at which the packaged processed cheese is cooled is another key influence on its eventual texture.

Thus, block, sliced, or spreadable processed cheese will differ in a number of ways, particularly the level of intact casein present (determined by the cheese selected for processing), the emulsifying salts chosen (and their influence on the pH of the final product, in particular, with products of higher pH being typically softer and more spreadable), and the amount of water added (this being higher, as might be expected, for softer and more spreadable products).

In conclusion, cheese is an ancient dairy product, which probably arose by accident but became popular as a means of preserving the nutritional quality of milk by applying several classical food preservation techniques simultaneously (dehydration, acidification, and salting). The general principles for the production of most types of cheese are relatively common (coagulation, fermentation, separation of curds, and ripening) but small differences in exact procedures, the milk (and its treatment), and the extent and temperature of ripening result in a great diversity of different varieties.

# 7
# Fermented Milks

Globally, there are a large variety of fermented or cultured milk products, probably because they are relatively easy to produce. Many countries produce unique indigenous fermented milks and, in this chapter, those that are produced worldwide, like yoghurt, cultured buttermilk, and sour cream, and some that have artisanal or limited production, like kefir, koumiss, and Scandinavian and Middle Eastern fermented milks, will be discussed.

The principles used in making fermented milks are very simple. Generally, milk, with or without increased solids, is pasteurized, cooled to the incubation temperature, a starter culture added, and the mixture incubated for several hours until the final desired level of acidity or pH is reached, after which the product may be packaged and refrigerated at 4°C. The reason for the increased solids is to produce a thicker product and prevent the expulsion of whey (syneresis). Like cheese-making, the key biochemical reaction is the production of lactic acid from the milk sugar, lactose, by the starter bacteria. This results in a decrease in the pH and a consequent increase in the shelf-life of the product. The starter bacteria typically used are thermophilic or mesophilic lactic acid bacteria (LAB) and include *Lactococcus*, *Lactobacillus*, *Streptococcus*, and *Leuconostoc* spp. (Chapter 4). Other compounds may also be produced from lactose, e.g., acetaldehyde (around 20 mg/L), which is responsible for the green apple or nutty flavor of yoghurt, and ethanol, produced by an alcoholic fermentation of lactose in the case of kefir and koumiss. Diacetyl, an important flavor compound of cultured buttermilk, is not produced from lactose but from citric acid, by strains of *Lactococcus lactis* and *Leuconostoc* spp.

Where fermented milks originated is not clear, but there is some evidence that kefir originated in the Caucasus Mountains, which link the Black and Caspian Seas, about 6000BC. It is easy to understand how they might have originated if one considers that, in principle, raw milk, contaminated accidentally with lactic acid bacteria and held at 25–45°C, coagulates and gives rise to a fermented milk. One of the major advantages of fermented products is their increased shelf-life, because of the low pH of 4.5 attained, due to the production of lactic acid.

*From Farm to Table.* Alan Kelly, Patrick Fox, and Tim Cogan, Oxford University Press.
 DOI: 10.1093/9780197581025.003.0007

## Yoghurt

The word yoghurt is derived from the Turkish word *yoğurt*, meaning to thicken. Where it originated is not clear, but it may have been in ancient Mesopotamia, around modern Iraq, between the Tigris and Euphrates rivers, where sheep and goats were first domesticated. The fact that yogurt is derived from a Turkish word also suggests that it may have originated in Turkey, perhaps in the province of Anatolia, where cattle were first domesticated around 4000 BC. These regions are very warm, and it would be easy to make a yoghurt from raw milk contaminated with the correct lactic acid bacteria if it was held at a temperature conducive to the growth of the LAB.

Yoghurt can be made from skim or full-fat milk, usually from cows, although sheep, goats, and buffalo milk can also be used. In the case of full-fat cow milk, the milk is normally homogenized at 20–25 MPa at 50–60°C to prevent the fat globules from rising during the fermentation (i.e., creaming) and to ensure a uniform distribution of the fat. The milk is also pasteurized; both homogenization and pasteurization are done as an integrated step, otherwise the milk would rapidly go rancid due to the activity of the lipase in the milk on the milk fat, producing fatty acids (and also glycerol and mono- and di-glycerides), many of which have a rancid taste; the lipase is inactivated by pasteurization. Homogenization improves the stability and viscosity of fermented milks, even those with a low fat content. Skim milk powder may be added to the milk before heat treatment to increase the milk solids level by 2%–3% and, after dissolution, the mixture is heat-treated at 90–95°C for 5 minutes before cooling to 42–43°C and adding a thermophilic starter (see Chapter 4). This high heat treatment results in denaturation of about 80% of the whey proteins, particularly β-lactoglobulin, which interacts with the casein and helps to give the yoghurt a firmer "body" and a product less prone to syneresis and the release of whey.

Stabilizers, added at low levels to the milk before fermentation, help to improve the body of yoghurt by preventing syneresis. These molecules, called hydrocolloids, are excellent at binding water, and when they do so they greatly increase the viscosity of the system in which they are found. Both the high heat treatment of the milk and the addition of stabilizers help to prevent undesirable syneresis in the yoghurt and, in addition, some stabilizers, such as pectin, can directly interact with casein, building complex and stable gel structures. Many commercial yoghurts today contain individual stabilizers or blends of these, which may have plant (starch, guar gum, locust bean gum), fruit (pectin), seaweed (carrageenan, agar), or microbial (xanthan gum) origins, or may be chemically modified to improve and tailor their properties (modified starches).

The starter used to make yoghurt is classically a mixture of two different lactic acid bacteria, a rod, *Lactobacillus delbrueckii* subsp. *bulgaricus* and a coccus,

*Streptococcus thermophilus*. Although both grow in milk by themselves, they grow better together, as each of them produces compounds that stimulate the growth of the other one (see Chapter 4). This property of mutual help with growth is called symbiosis. A photomicrograph of yoghurt showing the growing rod and coccus is shown in Figure 7.1.

The mixture of milk and culture is held at ~43°C for 4–6 hours, until the desired pH of 4.5 is achieved, and then cooled. Incubation at lower temperatures for a longer time is also commonly used today. This pH is the isoelectric point of casein, i.e., the pH at which the casein is least soluble and all of it has gelled. The low pH also increases the shelf-life of yoghurt.

There are two ways in which the fermentation is performed, giving rise to quite different end products. In one, the milk and culture are added to the yoghurt

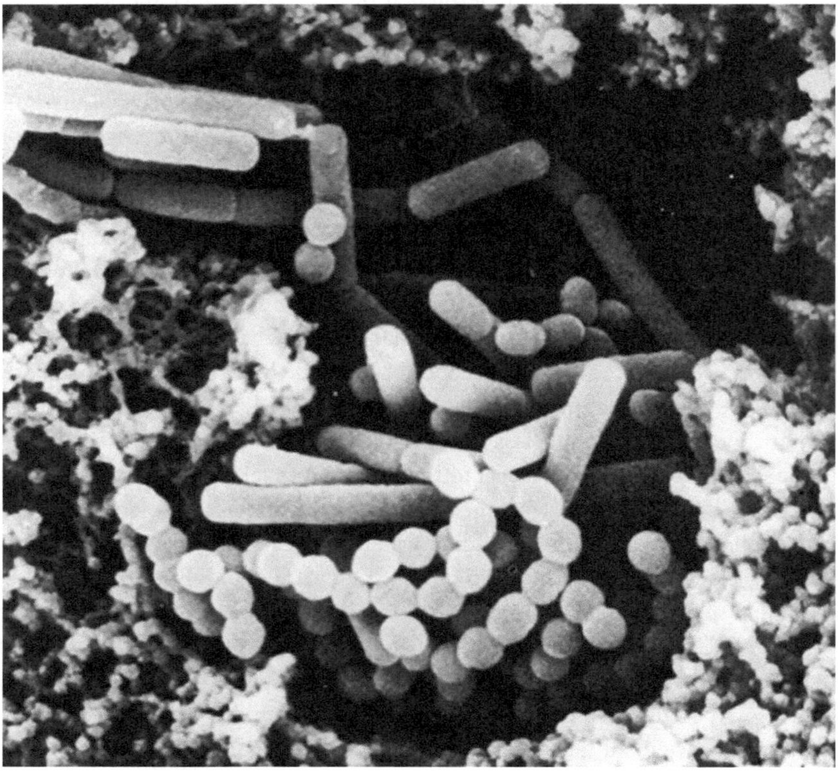

**Figure 7.1** Two distinct types of bacteria, rod-shaped *Lactobacillus delbrueckii* subsp. *bulgaricus* and coccal-shaped *Streptococcus thermophilus* growing in close proximity in a yoghurt gel (from Botazzi, V. 2002. . . .siamo piccolo e molto bravi in biochemical: produciamo lo yoghurt, *Industria Latte*, 38, 1–2, with permission).

pot or other container, the lid applied, and fermentation then proceeds. In this case, the gel is undisturbed until the lid is removed before eating, and a smooth firm gel is produced—this is called set yoghurt. In the other method, fermentation takes place in large bulk tanks and, when the final pH has been reached, the yoghurt is filled into its final package. In this case, the yoghurt, called stirred yoghurt, is more fluid. This system also allows fruit or other additions to be mixed in at the end of the fermentation, or yoghurt to be layered onto a fruit or other (e.g., biscuit) base before the package is sealed.

Other types of yoghurt are also produced, including fruit-flavored yoghurt and Greek-style yoghurt. In the production of fruit-flavored yoghurt, the yoghurt is cooled to 15–22°C after manufacture, sugar and fruit are added to about 10% of the total volume, and the product packaged. The fruit can be natural, i.e., unsweetened, or sweetened by adding 50–55% sugar to it before adding the mixture to the yoghurt. Both sugar and fruit must be heat-treated before addition to the yoghurt since they, especially the fruit, can be a potent source of yeast, which will produce $CO_2$ from the sugar, resulting in gas and off-flavors in the product. In the production of stirred yoghurt, an additional heat treatment (72–75°C for a few seconds before cooling) is sometimes given to the yoghurt after fermentation, followed by aseptic (sterile) packaging, which also increases the shelf-life. The heat treatment inactivates the starter culture and any contaminants that might be present, e.g., yeast and molds, and the aseptic packaging prevents recontamination.

Greek or Greek-style yoghurt is very viscous. This is accomplished by straining the yoghurt after fermentation to remove some whey, thus increasing the solids level and therefore its consistency. The solids level can also be increased by membrane filtration of the milk before adding the starter.

Yogurt is a popular snack or dessert product in many countries today, often combined with fruits, nuts, biscuit, or other meal components for consumption. Probiotic bacteria are also often added to yoghurt to improve its health benefits (as discussed below).

## Cultured buttermilk

Traditionally, buttermilk was a by-product of either lactic or sweet cream butter manufacture and contained only low levels of milk solids. The buttermilk produced from lactic butter manufacture is more like cultured buttermilk than that from sweet cream buttermilk.

Cultured buttermilk is a particularly important product in the United States. In Ireland, it is commonly used to make homemade brown soda bread—a traditional form of bread for which the $CO_2$ necessary to raise the bread is produced during cooking by the chemical breakdown of added sodium bicarbonate by the lactic acid present in the buttermilk, rather than by fermentation. The commercial manufacture of cultured buttermilk involves 3 steps: (1) separation of whole milk into skim milk and cream streams, (2) pasteurization of the skim milk and cooling it to 21°C and (3) inoculating it with a DL mesophilic culture and incubating for 12–16 hours until it coagulates, due to fermentation of lactose to lactic acid. The pH will reach 4.5 in this time, and the product is then ready for use.

Gas ($CO_2$), acetate, and diacetyl are also produced due to the metabolic activity of the leuconostoc and some strains of lactococci in the starter culture on the citrate present in the milk. The $CO_2$ gives a small amount of effervescence to the buttermilk and the diacetyl a lactic butter flavor. According to international definitions,[1] cultured buttermilk should have an acidity of at least 0.6%, calculated as lactic acid, be fermented with a DL culture, and contain at least 10 million ($10^7$) colony forming units (cfu) per ml at the point of sale.

## Sour cream

Essentially, sour cream is made in the same way as cultured buttermilk. The cream is first standardized to a minimum of 18% fat and a stabilizer, e.g., a polysaccharide, such as modified food starch, guar gum, carrageenan, or gelatin is added, which gives a smoother, more viscous body to the final product and reduces syneresis of whey. Sodium citrate may also be added to increase the buttery flavor, which is due to diacetyl production. The cream is heated to 60°C, homogenized to reduce the size of the fat globules, and pasteurized at 85°C for 30 min. After cooling to 21°C, the cream is inoculated with 1–2% of a DL culture (see Chapter 4) and incubated for 18 hours. During incubation, lactic acid is produced from the lactose in the milk, which reduces the pH from 6.6 to 4.6, and the $Cit^+$ bacteria in the starter culture produce diacetyl, an important flavor component of sour cream, from the citrate in the milk. The sour cream is then packaged into the final container; once cooled, it is ready for sale.

## Kefir

Kefir is a slightly effervescent fermented milk with an alcoholic flavor; its production involves both an alcoholic and a lactic fermentation by yeast and LAB, respectively. One of the unusual things about kefir is that the starter used in its production is a symbiotic mixture of lactococci, lactobacilli, leuconostocs, acetic acid bacteria, and yeast held together in a polysaccharide matrix, called kefiran.

This structure is called a kefir grain (Figure 7.2). It looks like a cauliflower floret and is unusual in being visible, recoverable, and continuously reusable, simply by straining the fermented kefir and washing the grains in cold water before using them as a starter for a new batch. As far as we are aware, it has not been possible to reproduce a kefir grain from the bacteria and yeast that are found in it. They can be bought from any culture supply laboratory and, in Ireland, they are often used in the production of buttermilk used for making soda bread in the home and for this reason are sometimes called buttermilk plants!

Various lactic acid bacteria, acetic acid bacteria, and yeast have been identified in kefirs of different origins. The bacterial component includes *Lc. lactis*, *Leuc. mesenteroides*, *Lb. kefiri*, *Lb. kefiranofaciens*, *Lb. parakefiri*, *Lb. helveticus*, *Lb. parabuchneri*, *Lb. amilovorus*, *Lb. crispatus*, *Lb. buchneri*, and *Acetobacter pasteurianus*. More than 23 different species of yeast have been isolated from kefir grains; the predominant species are *Saccharomyces cerevisiae*, *S. unisporus*, *Candida kefyr*, and *Kluyveromyces marxianus* subsp. *marxianus*.[2] Not all of the bacteria and yeast are found in every sample of kefir, and the role of each of these microorganisms in the production and flavor of kefir has not been determined.

Curiously, the composition of the grains does not mirror the composition of the fermented kefir milk, in that *Lc. lactis* is the main organism found in the kefir milk while it is only a minor component of the grain itself, which is dominated by

**Figure 7.2** The structure of a typical kefir grain, around 2 cm in width.

lactobacilli, which are also found in the kefir milk but at relatively low levels.[3] This suggests that the lactococci are present on the outside of the grain and thus are easily accessible for the fermentation. In the final product, lactococci number $10^8$–$10^9$ cfu/ml, followed by leuconostoc at $10^7$–$10^8$ cfu/ml, yeasts at $10^6$–$10^7$cfu/ml, and acetic acid bacteria and lactobacilli at about $10^6$ cfu/ml.[4] The presence of acetic acid bacteria in the grain is unusual in a fermentation, since one of their functions is to oxidize ethyl alcohol to acetic acid, which requires oxygen, even though the lactic and alcoholic fermentations are anaerobic (and so do not involve oxygen).

It is easy to make kefir. All one has to do is pasteurize the milk, cool it to about 21°C, add about 1% kefir grains and incubate for 24–48 hours, during which the pH will decrease to about 4.5 and the milk will coagulate. At this stage, the kefir grains can be recovered for reuse. Of course, they can also be purchased from starter culture suppliers, but it is frequently claimed that the best way to make kefir is by using the traditional grains.

## Koumiss

Koumiss (also called kumys, kumis, qynyz, airag, or chigee) is made from mare's milk by nomads living in tents, called yurts, on the steppe of Mongolia and Inner Mongolia, an autonomous region of northwest China, but also in Kazakhstan, Kyrgyzstan, and parts of Russia. The nomads were described by Herodotus (c. 484–425BC) as noble milkers of mares! Koumiss is thought to date from 2000BC, although the earliest record in Chinese literature dates from the Han dynasty (202BC–220 AD). Koumiss has been reported to have medicinal properties, healing intestinal dyspepsia, hypertension, and dyslipidemia.

Like kefir, the production of koumiss involves both lactic and alcoholic fermentations. The composition of a significant number of samples of koumiss was compared in a recent study,[5] and the pH, lactic acid, and alcohol contents and the numbers of LAB and yeast present in the product showed significant variation, as did the coliform counts, indicating problems with hygiene. This is not surprising when one considers that koumiss is made from raw milk by nomads, and no starter cultures are used Instead the nomad depends on the LAB and yeast present in the milk as contaminants to ferment it, unless back-slopping is practiced, where a small amount of yesterday's koumiss is used as a starter culture. Some producers use a variation of back-slopping: the repeated use of a small, cloth bag containing dried, unmalted, boiled millet and small fruits as a starter, which have been first immersed in koumiss for several days before being dried and stored at 4°C. The millet and fruit adsorb the microorganisms present in the koumiss, and the bag is also contaminated by microorganisms from its repeated use. A possible role of the fruit could be as a source of yeasts for the alcoholic fermentation.

Mare lactations are short, lasting from June to September, and usually the foal stands beside the mare when she is being milked. Traditionally, the fermentation of the raw milk occurs in wooden casks, large animal skin bags, or porcelain urns, placed outside the yurt at around 20°C for 1–3 days, during which the milk is stirred for up to 30 minutes with a wooden stick to ensure even mixing. The koumiss is then bottled and stored at 4°C. The stirring during the fermentation is unusual in a product made with LAB since oxygen can inhibit the growth of many of them, which are essentially anaerobes. Instead, the producer relies on natural contaminants in the milk to produce lactic acid and ethanol.

The bacteria present in koumiss show considerable variation and include *Lb. helveticus*, *Lc. lactis*, *Lb plantarum*, *Lb. kefirofaciens*, and *Streptococcus parauberis*. *Lb. helveticus* is a thermophile with an optimum temperature of 45°C. It would grow slowly at the incubation temperature (20°C) used in koumiss production but can grow well under acidic conditions, which may be why it dominated all samples. Numerous yeasts were also found, including *Sac. cerevisiae*, *Sac. dairensis*, *Kluyveromysces marxianus*, and *Kasachstania unispora*. As the fermentation is natural, the microorganisms are contaminants from the raw milk and the utensils used in milking. Because of this, significant variation in the species present would be expected.

## Yakult

Yakult is a Japanese probiotic, fermented skim milk drink, first produced in 1935. A strain of *Lb. casei*, which was isolated originally from a human intestine, and which has since been reclassified as a strain of *Lb. paracasei* subsp. *paracasei*, is used as the starter culture. Not all lactobacilli are able to reach the intestine in a viable state, but the strain used in Yakult does, as it is resistant to both gastric juice and bile. It has been shown to improve bowel movements in constipated people and may also help to maintain a healthy intestinal microflora.

There is little published information on the details of Yakult manufacture. The incubation temperature is 37°C, and the fermentation takes about 16 hours. The retail product contains a low amount (about 3.7%) of milk solids and a high amount (about 14%) of added sucrose. Actimel is a similar product to Yakult but made by Danone in France.

## Scandinavian fermented milks

Scandinavian countries and Iceland have their own individual fermented milks. Like koumiss, many were initially homemade but have become commercialized because of their popularity.

Viili is a Finnish fermented milk with a pleasant sharp taste, a strong diacetyl flavor, and a ropy texture caused by an exopolysaccharide excreted during the exponential growth phase of some of the starter bacteria, which prevents the separation of whey from the product during storage.[6] This gives it a viscous, firm consistency. It is produced in 20 Finnish dairies, and total consumption is about 30 million liters per year. Traditionally DL mixed-strain starters are used in viili production; in addition, they contain a yeast, *Geotrichum candidum*, which grows on the surface of the product, consumes lactate, and reduces the acidity of the product. The yeast also consumes oxygen and produces $CO_2$, which leads to a slight carbonation of the product. The traditional product was made in the home using back-slopping with a small quantity of a previous product as starter, but commercial production has been practiced in recent years.

Langfil is a Swedish fermented milk, which is also made in Norway where it is called tattemilk. It is made in a similar way to cultured buttermilk, except that full-fat milk is used and some of the lactococci in the starter cultures secrete polysaccharides, which increases the viscosity of the product. Traditionally, langfil cultures were produced by dipping a piece of cloth (in Sweden) or birch twigs (in Norway) into a batch of product, allowing them to dry, and then inoculating a fresh batch by placing them in fresh milk. Modern commercial production involves inoculation with the starter and incubation until the pH reaches 4.6, after which the product is cooled. Langfil has a shelf-life of 10–15 days.

Ymer is a Danish fermented milk that was developed in 1937 and today is often produced from milk concentrated by ultrafiltration. The name Ymer derives from Ymir, a primordial being in Norse mythology. Heat-treated milk is cooled to 20–27°C, inoculated with an L type culture, and incubated until the pH reaches 4.5.

Skyr is a traditional Icelandic dairy product that is mentioned in Icelandic sagas dating back at least a thousand years. In contrast to the other Scandinavian products, skyr is made from skim milk using a yoghurt starter that also includes *Lb. casei*. Lactose-fermenting yeasts are also often isolated from home-made skyr. After manufacture, the product is thickened by removing some of the whey, either by straining through a linen cloth or by separation in a quarg separator, to give a product with 16–21% solids. In this regard, it is more like a fresh, unripened cheese, e.g., cottage cheese or quarg, than a fermented milk, except that the incubation temperature is higher (at 37–42°C) to promote the growth of the thermophilic culture. Whey proteins, recovered by ultrafiltration of the whey, are also sometimes added to the skim milk before heat treatment and cooling, to improve the strength of the gel, as they do in yoghurt production.[7]

## Middle Eastern and Indian fermented milks

Middle Eastern fermented milks can be divided into those with normal milk composition and solids level, e.g., zabady and laban, and more concentrated products containing 20%–40% milk solids, e.g., labneh and labaan zeer.[8] Zabady, a set-type yoghurt from Egypt, is usually made from partially skimmed buffalo milk, which has been boiled for a few minutes then cooled to 37–45°C before being inoculated with a small amount of a previous batch of product. It is then filled into pots, which are left uncovered until coagulation occurs, during which time the fat rises to the surface. The product contains predominantly *Sc. thermophilus* and *Lb. delbrueckii* subsp. *bulgaricus* but also non-lactic organisms, e.g., coliform, *Microbacterium lacticum* and yeasts, all of which are contaminants that probably gain entry to the product because the pots are left uncovered during the incubation step.

Labneh, a popular product in Egypt, Saudi Arabia, and Lebanon, is made from sheep, goat, or buffalo milk (the method of manufacture differs slightly in the different countries). In Egypt, yoghurt is stored overnight in a cool room; the following day, salt is added and mixed well, before the mixture is transferred to cheese cloth bags, which are then hung on racks for 12–24 hours to promote whey drainage (the salt acts as an inhibitor of bacterial growth). It is then packaged and stored at refrigeration temperature. In Saudi Arabia, the cheesecloth bags are pressed lightly to facilitate whey drainage, while, in Lebanon, no salt is added to the yoghurt. Several methods have been developed to overcome the time necessary for whey drainage and improve hygiene, including removal of whey by mechanical separation, culturing ultrafiltered milk, ultrafiltration of the fermented milk, or use of milk solids reconstituted to a solids level similar to that of traditional labneh.

In Asia, particularly India, Bhutan, Sikkim, and Nepal, dahi is an important dairy product produced from buffalo, cow, or goat milk, using a mesophilic culture for mild flavored dahi and a mesophilic culture plus *Lb. delbrueckii* subsp. *bulgaricus* for sour dahi.[9] Dahi is made in a similar manner to cultured buttermilk, except that the milk is full fat and is heat-treated to 95°C before cooling and inoculation.

## Probiotic products

Probiotics are defined by the Food and Agriculture Organization of the United Nations and the World Health Organization (FAO/WHO) as live microorganisms which when administered in adequate amounts confer a health benefit on the host. Yoghurt and other fermented and indeed non-fermented drinks are often used as vehicles for carrying probiotic bacteria. The health benefits are many and varied and include, alleviation of lactose intolerance, prevention and reduction of diarrhea, treatment and prevention of allergies, reduction of blood cholesterol levels, inhibition of *Helicobacter pylori* (a cause of stomach ulcers), prevention of inflammatory bowel disease and modulation of the immune response.[10] Whole kefir, as well as specific fractions and individual organisms isolated from it, are reported to provide a multitude of positive effects when consumed. These range from improved metabolism of cholesterol, reduced blood pressure, antimicrobial activity, suppression of tumor growth, improved wound healing and modulation of the immune system and the intestinal microbiome, and even the potential alleviation of allergies, cancer, and asthma.[11]

The European Food Safety Authority has developed the Qualified Presumption of Safety (QPS) concept for selecting probiotics based on reasonable evidence. If the assessment concludes that a group of microorganisms does not raise safety concerns, the group is granted "QPS status." To be granted QPS status, a microorganism must meet the following criteria:

- its taxonomic identity must be well defined;
- the available body of knowledge must be sufficient to establish its safety;
- the lack of pathogenic properties must be established and substantiated;
- its intended use must be clearly described.

Probiotic therapy is not a new idea; it dates back more than 100 years to Elie Metchnikoff (an Ukrainian Noble prize winner for his contribution to immunology), who, in 1907, suggested that Bulgarian peasants lived long lives—many of them to over 100 years—because they consumed so much yoghurt. In the 1930s, a Japanese physician, Minoru Shirota, suggested that the right mix of bacteria in the gut could prevent intestinal disease; he developed the probiotic drink Yakult (described above).

Probiotic bacteria are colloquially called "friendly" bacteria and are generally strains of lactobacilli and bifidobacteria, commonly found in the intestine. They are selected on the basis of several criteria, e.g., being normal inhabitants of the intestine, non-pathogenic, capable of colonizing the intestine, lack of virulence and antibiotic resistance factors, ability to tolerate gastric juices and bile, ability to adhere to the intestinal mucosal surface, and shown to have the desired effect in clinical trials. Generally, high levels, around $10^7$–$10^8$ cells, need to be taken to have the desired effect.

# 8
# Butter

Butter is a dairy product with a very long history. It was considered to have medicinal properties by the Romans and has been found buried in bogs in many European countries; some of these bog butters date back thousands of years. They may have been buried in the bogs to preserve them (due to the low pH and temperature of the bogs) in times of plenty for use in times of need or to protect them from raiders. Alternatively, these may simply be discoveries of containers lost centuries ago. A container of bog butter from the late medieval period (1150–1550AD) in Ireland is shown in Figure 8.1.

A study published in 2007 by Irish researchers analyzed samples of bog butter in national museums and collections, dating from as long ago as the Iron Age (400BC—500AD) up to the 17th century. The samples (which contained no salt and very little proteinaceous material) had dried to very low moisture contents, but had recognizable fatty acid profiles, albeit broken down to very simple end products.[1]

In the 18th and early 19th centuries, butter made in farmhouses was Ireland's most important agricultural export. The reasons for this may be its ease of manufacture and its ability to store well. All one had to do was to first store the milk in suitable vessels for several hours. Once the cream rose to the top, it could be collected by skimming into another vessel and then churned, probably in a dash churn—a barrel with some sort of paddle—until butter and buttermilk appeared. Two traditional butter churns that were used in farmhouses or homes are shown in Figure 8.2. In the Iraqi Museum, Baghdad, there is a frieze dating to c. 2500BC showing cows being milked and what one could construe as butter being made.

Butter consists of ~83% fat, 16% moisture (legal maximum), and sometimes ~1.5% salt (NaCl). It is essentially a water-in-oil emulsion with the moisture distributed as droplets discontinuously in the fat, which is the opposite of the oil-in-water emulsion state of cream. Conversion of cream to butter essentially involves inversion of the emulsion, to turn oil (milk fat) from the dispersed phase to the continuous bulk phase. It is sometimes said that butter is actually not a true emulsion but rather a pseudo-emulsion, as the water droplets are entrapped but not surrounded by an emulsifying layer.

*From Farm to Table.* Alan Kelly, Patrick Fox, and Tim Cogan, Oxford University Press.
 DOI: 10.1093/9780197581025.003.0008

**Figure 8.1** Bog butter from the late medieval period in Ireland in a circular, hollowed-out wooden container, on display in the Butter Museum, Cork, Ireland, on loan from the National Museum of Ireland (from Roland Paschhoff, with permission).

In making butter, cream is churned until it "breaks," forming butter granules and buttermilk. The granules are then washed free of the buttermilk and "worked" until the moisture is evenly distributed throughout the butter.

Two types of butter are made, sweet cream and lactic, each of which may be salted or unsalted (lactic butter is normally unsalted because salt, low pH, and traces of copper promote oxidation of the fatty acids). Until the end of the 19th

(a) (b)

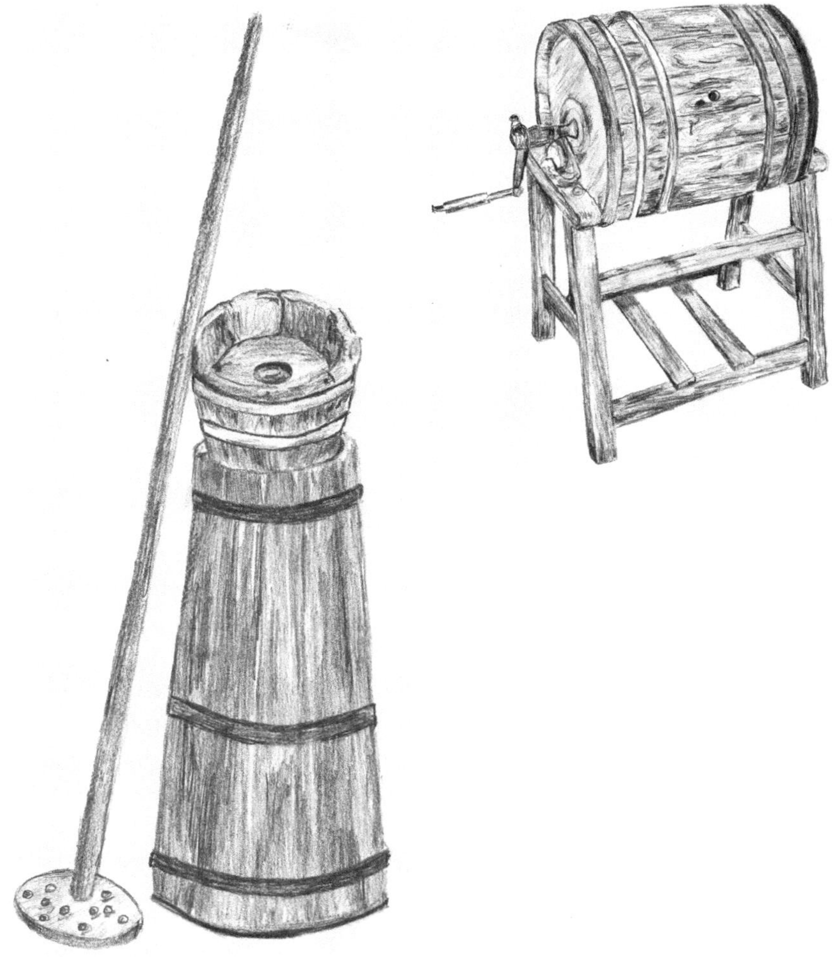

**Figure 8.2** Two common types of small-scale traditional butter churns, (a) the dash churn where agitation is provided by the vertical paddle, and (b) the rotary churn, which is rotated by turning of the handle on the left (images thanks to Anne Cahalane).

century, the production of butter was, in all countries, farm-based, and salted lactic butter was probably the most commonly made type. This was due to the fact that milk was not refrigerated before churning and a lactic fermentation of the cream occurred due to growth of contaminating lactic acid bacteria.

Today, unsalted lactic butter (with a pH of 4.9–5.3) is the butter most commonly made in continental Europe, while salted sweet cream butter (pH 6.6) predominates in most English-speaking countries. Just when sweet cream butter became the common butter in English-speaking countries is not clear. In textbooks published in the United States, United Kingdom, and Germany in the 19th and early 20th centuries, cream, either ripened naturally (i.e., without addition of a starter culture) or with the addition of a starter culture, is described as being used to make butter.

Toward the end of the 19th century, a number of major developments of relevance to butter-making occurred:

- the establishment of creameries in many countries where a number of farmers pooled their cream for larger-scale butter manufacture;
- the development of the centrifugal separator, which could separate cream from milk mechanically and efficiently, by Gustaf de Laval in Sweden in 1878;
- the development of pasteurization of milk by Franz von Soxhlet in 1879;
- research around 1890 in the United States, Denmark, and Germany independently showing that the ripening of cream was due to the growth of lactic acid bacteria.

These developments laid the foundation for the commercialization of butter-making and starter cultures. These cultures soon became commercially available, and their use expanded rapidly, e.g., in 1897, 802 out of 866 butter-making plants in Denmark were using commercial starters. In 1919, it was shown that highly flavored butter was due to the use of starters consisting of both lactococci and leuconostocs (L cultures, see Chapter 4). The former produced the lactic acid necessary for fermentation from the lactose in the cream and the latter produced the flavor compound, diacetyl, from the citrate present in the cream.

## Principles of butter-making

In making butter, only two ingredients are required, cream and air, along with significant physical force. Butter can be made from whole milk, but, in practice, it is considerably easier to start with cream in which the fat content has been greatly increased, originally by gravity settling, but today much more quickly using mechanical separation, as discussed in Chapter 5.

Making butter is thus a relatively simple process (Figure 8.3); all that is needed is cream and a suitable container with a lid. This could be a glass or plastic jar into which cream, maybe a little warmer than refrigeration temperature (say

(a) (b) (c) (d)

**Figure 8.3** Stages in the production of butter, showing (a) cream, (b) whipped cream that has changed in volume and firmness, (c) separated mass of butter grains suspended in buttermilk, and (d) butter after draining off the buttermilk and working the butter (images thanks to David Waldron).

10–12°C), is placed, leaving plenty of room for air (maybe one-third to half the container volume). Then, the container is vigorously shaken, mixing the cream and air for several minutes. In the past (and occasionally still today), the container used would have been a wooden barrel and the agitation was supplied either by rotating the barrel in a frame using a handle or using mechanical plungers in an up-and-down movement.

The progress in converting cream into butter can, interestingly, be monitored by sound throughout the process. When agitating the cream, the first thing that will happen after a few minutes of mixing is that the initial sound of liquid sloshing around will stop, and the cream will have solidified and greatly increased in volume. At this point, the cream is whipped, and could be used as a dessert topping. If agitation is continued, the sound of liquid moving in the container will resume as the emulsion breaks, and after that the knocking of particles against the inside walls of the container becomes audible. Small yellow "nuggets" called butter grains will be apparent in the cream, and further agitation will cause these to grow in size and separate from the surrounding liquid, called buttermilk.

As the grains become more defined, the buttermilk is drained off, and further agitation (called "working") fuses the individual grains into a homogeneous mass, which eventually results in a smooth butter that can be removed. If salt is to be added, this is usually done after the buttermilk has been drained but before the working stops (to allow it to be evenly distributed). The butter is then refrigerated to cool and solidify further.

Thus, butter is relatively easy to make, which is probably why this has been a common way of preserving the most energy-dense portion of the milk (i.e., fat) into a stable concentrated form. However, the product made by the method described here has one major deficiency; if taken from the fridge and applied directly onto bread, it will be extremely hard and difficult to spread evenly. It is thus easy to make butter, but not necessarily easy to make good spreadable butter.

To understand how to make spreadable butter, and thereby make a product more suited for its expected use by consumers, it is necessary to think more deeply about the science of what is going on in the process.

## The solid-liquid nature of milk fat

As discussed in Chapter 2, the fat in milk occurs in the form of emulsified droplets, or globules, that float suspended in the milk serum or skimmed milk. To produce butter, it is necessary to destabilize these globules and release the fat, but first it is important to consider how the properties of milk fat affect one of the key properties of the final butter, which is its hardness.

Butter is used mainly for spreading on bread, biscuits, or crackers (Figure 8.4). To work well in these roles it requires a texture that is neither so soft that it would flow like a liquid, creating quite a mess, nor so hard that it would be impossible to spread and applying force using a knife would merely transmit the force to the bread, fracturing and tearing it.

One of the principal factors that affects the spreadability of butter is temperature: straight from the fridge, butter is much more solid and less spreadable than if it has been left stand at room temperature. Why is this the case?

**Figure 8.4** A typical use for butter today, melting nicely on warm toast (image thanks to David Waldron).

As discussed earlier in this book, milk fat comprises triglycerides, made of fatty acids bound to glycerol. Such triglycerides can be solid or liquid at any given temperature depending on whether it is above or below their melting point, which, in turn, depends on factors like the number of carbon atoms present in the fatty acids and whether they are saturated or unsaturated. Whereas water melts at 0°C, in the case of milk fat, melting occurs not at one specific temperature but across a relatively wide temperature range, as each individual triglyceride has its specific melting point. All milk fat is solid at around −40°C and liquid above 40°C, and the proportion that is liquid increases gradually as the temperature increases between these values. When butter is taken out of the fridge, quite a high proportion is in crystalline form and it is hard, but as it warms up, progressively more triglycerides reach their melting point and liquify. At room temperature, more of the butter is thus liquid and the product is more fluid and spreadable.

Compare this behavior to that of most vegetable oils, such as olive oil or sunflower oil, which are sold as liquids in bottles. These oils contain a high proportion of unsaturated fatty acids which, due to the presence of carbon-carbon double bonds, have low melting points, and so the oil is fluid at room temperature (and can be even in the fridge, in the case of salad dressing or mayonnaise containing such oils). Animal fats such as butter contain a much higher proportion of saturated fats, and hence are at least partially solid under similar conditions.

How then can the key property of butter for consumers, i.e., spreadability, be controlled and optimized to give a product that is neither too hard nor too soft to use? There are two main strategies to do this.

The first relates to the origin of the fatty acids, many of which are derived from the cows' diet. There is thus a relationship between the cows' feed and the fat composition in their milk and, to a certain extent, the composition and properties of milk fat can be "programmed" by manipulating the cows' diet. For example, the more unsaturated oils a cow consumes, the more these will be represented in the milk fat, and the softer the fat will be. Bacteria in the cows' rumen can hydrogenate some dietary fats, converting unsaturated to saturated fats, but this can be avoided by providing fats in a protected form, which decreases their susceptibility to this reaction.

In Ireland and New Zealand, the relationship between diet and fat composition is very apparent, as cows consume grass extensively in the spring, summer, and autumn seasons, leading to the presence of carotenoids in milk fat, which gives butter made from that milk a characteristic golden/yellow color.

Thus, the composition, at a fatty acid and triglyceride level, is largely predetermined by the time the milk leaves the farm, due to factors like feeding practices, inclusion of grass in the feed, weather, and any supplements or concentrates the cows have been fed. For this reason, the hardness and spreadability of butter are likewise to a large extent predetermined by the time the milk leaves the farm, and these key properties of butter vary over a year, between years, and even between farms, in a manner that is not entirely controllable by the butter manufacturer.

While the hardness of butter is greatly determined at farm level, there is still an opportunity to influence it in the factory. This is because, while the proportion of fat supplied that will be liquid or solid at various temperatures is fixed at this point (by the fatty acid composition), the nature and structure of the crystalline fat is not.

Fats form crystals when they solidify. However, there are many types of crystals: some are large, some are small, and they have different shapes; some are pure, containing one type of triglyceride only, and some are mixtures of different triglycerides that have become entangled in each other as they crystallized.

Thus, while it is not possible to control how much of the milk fat will be solid (crystalline) in the factory, it is possible to control the nature and behavior of the crystals that form. Two samples of butter with similar levels of solid fat can have very different textures if their crystals are of a different size, shape, and purity. The way in which the nature of these crystals is controlled and tailored is linked to the handling and cooling of the cream in butter-making.

Milk intended for butter manufacture is first separated into skim milk and cream, as making butter from milk without such separation would be a lengthy and difficult task (even farmhouse butter-makers let the milk stand to recover cream before churning). The cream is then pasteurized, typically at 80–85°C for 15–30 seconds, for reasons discussed later; at such high temperatures, all the milk fat liquifies. The cream is next cooled to a temperature around 10–12°C, if going straight into the butter-making process, or otherwise refrigerated. In cooling from the hot liquid state, many of the triglycerides in the cream pass through their melting point and crystallize, and the key factor in terms of the final butter structure is the temperature profile through which they reach this point.

Picture freezing water. Just as no two snowflakes have the same shape, ice crystals can have a wide range of sizes, which depend directly on the rate at which the water freezes. The size of ice crystals is an important factor when considering the nature of frozen food, as water expands when it freezes, and the crystals formed can be very jagged and rough, damaging delicate structures in which they form. If something is frozen extremely quickly, e.g., by immersing it in liquid nitrogen at −196°C, the ice crystals that form are very small and do very little damage to the product; sensitive materials such as biological materials (e.g., bacteria and human organs) are frozen in this way to avoid damaging them by ice crystals.

Likewise, the size of fat crystals that form in the cream (and thereby the structure of the butter made from it) can be manipulated by controlling the rate at which the cream is cooled and crystallizes after pasteurization. Fast cooling results in a lot of small crystals, while slow cooling produces larger and fewer crystals; the combination of crystal size and surface area (which relates to their ability to adsorb the liquid fat present) greatly influence the hardness of the butter.

In one common process, butter-makers rapidly cool cream to form lots of small crystals and then rewarm it to a temperature of 20–25°C, to melt some of the fat and allow a more desirable crystal structure to form. However, at certain times of the year, a progressive cooling process without this warming step might be preferable.

How does the butter-maker decide which method to use to cool the cream? One key factor is season, with summer profiles (when cows are eating a lot of grass) often being different to those used in winter. The milk fat may also be tested to determine how much saturated and unsaturated fat is present, so that a cooling program can be chosen accordingly. A simple test is to add iodine and measure the reaction that takes place; iodine binds to (and hence is taken up by) double bonds, and so, the higher the “iodine number or value” of a sample of milk fat, the more double bonds and the more unsaturated fat there is present. Should the iodine value of the fat in a sample of milk be in a certain range, one type of cooling profile will then be used, while if a different iodine value is found a different profile would be selected.

In this way, what might appear to be a simple consideration, cooling the cream after pasteurization, turns out to be one of the key steps in which the properties of the final product can be controlled, involving a process (called "tempering") that takes several hours or is allowed to proceed overnight. Tempering actually represents the main way in which butter manufacturers compensate for variations in fat composition (specifically, relative proportions of saturated and unsaturated fatty acids) due to seasonal changes in cows' diet.

## Modern butter-making

Today, cream is separated from the milk by centrifugal force and the cream, containing about 40% fat is pasteurized. The heating process applied is more severe (80°C for 15–30 seconds) than traditional pasteurization (72–74°C for 15–30 seconds), for a number of reasons. Firstly, cream contains a high level of fat, which acts as a barrier to heat transfer (as fat is an insulator), and so to ensure adequate inactivation of undesirable microorganisms a higher temperature is used than for milk.

Secondly, milk contains an enzyme (lipase) that acts on milk triglycerides producing free fatty acids, which cause development of rancid and undesirable off-flavors. In milk, this happens only to a very limited extent because a barrier (the milk fat globule membrane) separates the enzyme from its target (substrate) and so the reaction cannot take place. Making butter, however, depends on removing that barrier in the process of releasing the milk fat and would result in rancidity if the milk was not heated sufficiently to ensure the enzyme is inactivated, requiring temperatures slightly higher than those used for traditional milk pasteurization.

The cream is then cooled and tempered, as discussed above, before it is churned and worked in a traditional churn or continuously in a butter-making machine. In some smaller butter plants, large batch churns, which may be cylindrical, cuboidal, or tetrahedral, are used. Into these metal churns, cream is loaded—only partially filled to allow for significant amounts of air to be present—where fast mechanical rotation results in churning and "breaking" of the cream. The churn is opened to remove buttermilk and closed again for further agitation and working, perhaps interrupted for the addition of salt if desired, until butter is formed and removed from the churn. This is referred to as a batch process, as it starts with the loading of cream into the churn and ends with the removal of butter some hours later. A batch churn is shown schematically in Figure 8.5.

**Figure 8.5** A typical batch butter churn, with the hatch through which cream is added and butter and buttermilk removed open. The churn is rotated around the axis in the middle by a mechanical motor (image thanks to Anne Cahalane).

In larger-scale butter plants, a continuous butter-making machine is typically used. In such a system, cream continuously enters the machine, and buttermilk and butter are (separately) continuously removed throughout a day of production. This is called the Fritz process, and the machine used consists of a number of sections (as shown in Figure 8.6). In the first section, shaped like a horizontal cylinder, the cream is rotated and agitated at high speed in the presence of air, which causes churning, and a mixture of butter grains and buttermilk leaves this section vertically (by gravity) into the next section. This is a tube, angled upward, fitted with a rotating auger screw that pushes the butter grains forward,

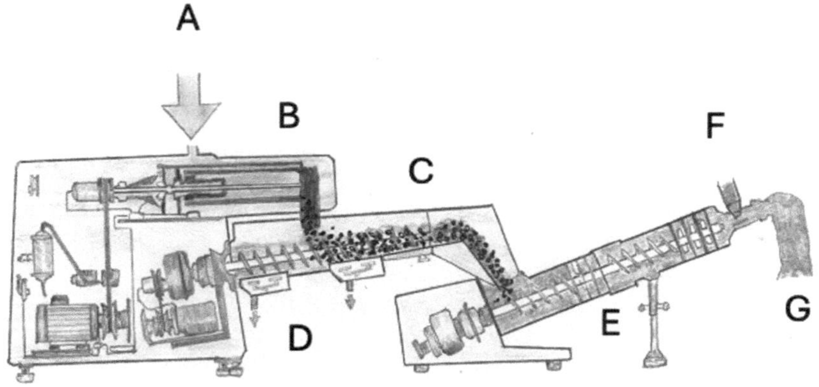


**Figure 8.6** Schematic diagram of a continuous butter-making machine, for what is known as the Fritz process. Cream enters at (A) and is rapidly churned in the first stage (B), leading to a mixture of butter grains and buttermilk, which pass into an upward-sloping chamber (C), where a rotating auger pushes the grains upward and forward while the buttermilk drains by gravity (D). The grains are then pushed by another auger through a series of perforated plates (E) to break down their structure, before salt is added (if necessary) and the moisture level adjusted (F), leading to butter exiting the machine (G) (image thanks to Anne Cahalane).

while the buttermilk flows by gravity in the opposite direction for removal. The separated butter grains then pass into another tubular section, where they are forced through perforated metal plates that break them down ("working") and make the butter mass more homogeneous; salt may also be injected (typically as a concentrated slurry) to be mixed in at the same time. At the end of this section, butter is extruded as a continuous mass, which can then be packaged and cooled.

During working, the moisture present (as a small amount of entrapped buttermilk) is distributed in a discontinuous way as very small droplets within the continuous fat phase, which are too small to sustain microbial growth. If the butter is salted and worked, the salt is evenly distributed within the moisture droplets. The salt is actually dissolved in the moisture phase of the butter, so a salt content of 1.8% in a butter of 16% moisture is an effective salt concentration of 11.3% in the moisture phase, which makes butter a very stable product microbiologically, as many microorganisms cannot grow at such high concentrations of salt.

Traditionally, once the cream had "broken" to form butter grains and buttermilk, and the buttermilk was removed. the grains were washed with cold water, but this is not common today because of the environmental problems associated with disposal of large quantities of wash water effluent.

## Lactic butter manufacture

Lactic butter is made in exactly the same way as sweet cream butter, except that the cream is first fermented with a DL starter culture (Chapter 4) until the pH reaches 5.5, at which point the temperature is reduced to 12–14°C and incubation continued until the pH reaches 4.6. The nutty flavor of lactic butter is due to the production of small amounts of diacetyl from citrate by the starter culture. Salt is normally not added to lactic butter because it promotes the development of rancid off-flavors, due to the oxidation of the double bonds in the fatty acids. In the past, the use of copper pipes promoted the oxidation, as some copper was stripped from the pipes by the milk or cream as it passed through them. As stainless-steel pipes are now more common, this defect is rarely encountered.

In 1978, a new process for making lactic butter was developed at the Netherlands Institute for Dairy Research (NIZO) and is now used extensively. In this method, the acid and flavor components are added directly to sweet (unfermented) butter grains, after draining off the buttermilk. The major advantage of this process is that sweet buttermilk is produced, a much more useful by-product economically than the sour buttermilk produced in the traditional method.

In the NIZO process, sweet cream is churned in the normal way and, after the cream has broken and the sweet buttermilk has been removed, two particular starter cultures, one of which has been acidified with lactic acid and aerated, are added to the washed butter grains and worked into the butter. This results in a butter with a pH below 5.3, a legal requirement for lactic butter made in The Netherlands, and which has a flavor and character very close to that of traditionally made lactic butter.

## Utilization of buttermilk

Buttermilk is the by-product of making butter and typically, for a given weight of cream, about half that weight of butter and the same weight of buttermilk are produced. Compared to skim milk, buttermilk has a slightly higher level of fat (perhaps 0.5% compared to 0.05%) but the key difference is the presence of the components of the milk fat globule membrane (mainly phospholipids and proteins) that were removed from the milk fat globules during churning. These give the buttermilk a number of unusual properties, including a rather strong and distinctive taste and a tendency to "cling" to glass containers in which it is held.

Buttermilk was widely regarded as a problematic and low-value dairy by-product and, while it is a popular drink in some countries in which lactic butter is made (such as The Netherlands), in many countries it was dried into powder for such uses as animal feed and baking or confectionary recipes, where it was a source of milk solids and strong dairy flavors.

In recent years, it has been recognized that the milk fat globule membrane (MFGM) components present include some biologically active molecules, and so buttermilk itself, or fractions recovered from it, have added value as health-promoting foods or ingredients. Claimed health benefits associated with the consumption of buttermilk include reduction in blood pressure and control of cholesterol levels, probably linked to the presence of MFGM-derived phospholipids.[2]

## Ghee and anhydrous milk fat

If butter is melted and held at a temperature above 50°C, it separates into milk fat and the aqueous phase of the butter, which, being denser, is the lower layer (such separation can be greatly accelerated by centrifugation). This can be drained off readily, to give a mass that is essentially pure milk fat, known as anhydrous milk fat (AMF). AMF may be used as a food ingredient (e.g., in chocolate, bakery products, or ice cream), as it is very stable (due to the lack of water) and confers a strong dairy flavor. It can also be recombined with skim milk (or reconstituted skim milk powder) to produce full-fat milk.

AMF can also be produced from cream, which has been separated to a higher fat content (75%) than the cream for butter-making (40% fat), then agitated to destabilize the emulsion, and centrifuged to separate the fat from the residual buttermilk.

AMF can also be processed to give milk fat fractions with different properties. For example, if it is melted (above 50°C) and then cooled and held at 21°C, all triglycerides with a melting point above 21°C will solidify. These solids (called the stearin fraction) can be separated (by filtration or centrifugation), and the milk fat that remains liquid (i.e., that with melting points below 21°C) is called the olein fraction. These two fractions have very different texture properties. The olein fraction, in particular, remains liquid at refrigeration temperatures, much like vegetable oil, and is a valuable ingredient in spreads, as will be discussed later.

A product closely related to AMF is ghee (from the Sanskrit word *ghrita*), which is a traditional Indian product and is produced like butter, but has a much lower moisture content as the milk fat is concentrated to almost AMF-like levels of purity. It can also be produced from butter by simmering (which helps to develop a stronger flavor) and then removing the separated molten fat. Ghee is referred to as a clarified butter and, often mixed with spices, is widely used in Indian cuisine. Lactic versions (called desi ghee) can be produced from ripened cream butter. Ghee is reported to be better than traditional butter for frying, as it has a higher smoke point, or temperature at which it begins to break down to undesirable odors.

## From butter to spreads

In many countries, it was typical, up to about 20 years ago, to see only butter on supermarket shelves as the main spread for use in the home on bread, with perhaps some margarine for baking purposes. Today, however, it is common to see a wide range of butter alternatives or substitutes, with differing fat contents, fat sources (like olive oil), and added ingredients, and claims such as the ability to reduce cholesterol levels due to the presence of plant sterols.

Why did these products emerge?

To consider the evolution of alternatives to butter, it is useful to start with the basic question of what problem these products were trying to solve, which reflects the objections of some consumers to butter:

- It is perceived as unhealthy due to its high content of saturated fat.
- It is expensive.
- It is user-unfriendly because it is not spreadable from the fridge.

The basic principle of the butter alternatives or "spreads" is that these three disadvantages (although the first one is somewhat controversial today) can be overcome by the replacement of some or all of the milk fat by cheaper, softer, perhaps healthier vegetable oils.

A simple spread could be produced by taking butter and blending it with vegetable oil, to reduce the average melting point of the fat present, and so make it more liquid and more spreadable at any temperature.

Other products avoid the butter process entirely, and formulate the product from scratch, blending oil-based and aqueous ingredients together to give a product that has the consistency and appearance of butter, but addresses the purported shortcomings listed above.

Vegetable oils have certain disadvantages as ingredients for these products, which add complexity to their use:

1. They are liquid, so a product that includes too much liquid oil is not spreadable but fully fluid. To overcome this hurdle, an industrial chemical process was developed in 1901, by a German chemist called Wilhelm Normann, to hydrogenate oils and thereby make them more solid and thus more suitable for use. The addition of hydrogen atoms removes the double bonds, making the fatty acids saturated, with consequently higher

melting points. However, concerns have arisen in recent years about the health implications of consuming hydrogenated oils, and specifically a by-product of the hydrogenation process, *trans* fatty acids. In these fatty acids, the double bond has not been removed, but rather the arrangement of the hydrogen atoms altered, and there have been some reports that this type of fatty acid is linked to elevated cholesterol levels in blood, and hence heart disease. There have also been reports of links to cancer and diabetes, though a European report on this concluded that the evidence was weak and inconclusive. Overall, such products are currently not in favor, and *trans* fats are banned in the United States, Brazil, and Canada.

2. Unsaturated fatty acids are susceptible to oxidation—oxygen atoms can react with the double bonds and yield off-flavors. For this reason, many spreads contain antioxidant compounds which "mop up" the products of the attack by oxygen and prevent oxidation-induced flavor deterioration. Indeed, the most common food ingredients added that have antioxidant functions are vitamins, like vitamin E, and so spreads are often enriched or fortified with vitamins both for their nutritional benefits and their role as protectors of oxidation.
3. Vegetable oils, compared to milk fat, can often be pale, and when present at a high level in a spread give it a pale white color. For this reason, spreads often contain the most natural source of butter-like color, a carotenoid such as β-carotene.

There is an additional complication in producing low-fat spreads; the less fat that is present, the more water will dominate the structure and the more likely it is that the product could be excessively fluid. In this case, the key goal is to add structure to the water droplets (as these products are still water-in-oil emulsions) by including ingredients, especially polysaccharides like alginate or gums that increase the viscosity of the aqueous phase and reduce its tendency to flow. In addition, the higher levels of water in low-fat spreads increase the risk of microbial growth. As increasing the level of the main preservative in butter (salt) would result in an undesirable flavor, unflavored microbial inhibitors, such as potassium sorbate, may be added to prevent microbial growth.

For these reasons, low fat spreads have a long and complex ingredient list, much more than butter (which is just cream and salt) and even full-fat spreads, and consumers today may make judgments on the desirability of such products by weighing up their concerns about fat intake, and specifically saturated fat intake, versus their perceptions of vegetable oils, preservatives, and long and complex ingredient lists. Because of these considerations, the popularity of such products is declining today.

# 9
# Concentrated and Dried Dairy Products

As water is one of the key agents through which undesirable changes occur in milk (or indeed any food product), by supporting microbial growth and chemical and enzymatic reactions, its removal or reduction can obviously greatly extend the shelf-life of products. The shelf-life of raw milk is measured in hours or days, that of pasteurized milk in weeks, and that of milk powder in years.

A number of dairy products rely for their production on the partial or almost total removal of water, i.e., concentrated and dried products. These will be the focus of this chapter. There are three key technologies involved—concentration, sterilization, and drying—which will be considered in turn, alongside the products they are used to create.

## Evaporated and sweetened condensed milk

Since the 19th century, sterilized milk in cans (produced by a process called retorting, because the equipment used to sterilize cans is called a retort) has been a common dairy product, valued for its extremely long shelf-life. The first tests on heat-sterilized milk were carried out around 1809, by a Frenchman, Nicholas Appert, who is regarded as the father of canned food. These tests were apparently unsuccessful, presumably because the milk coagulated. Around the same time, tests on heating mixtures of milk and sugar were reported in England, but the first successful commercial sweetened condensed milk product was produced in the United States by Gail Borden in 1856. He used sugar to preserve the product, which was more successful in producing long-life products than heat treatment of unsweetened concentrated milk. The first successful commercial production of milk concentrated by evaporation was in Illinois around 1885.

Today however, the far gentler heat treatment required to achieve the same degree of stability and safety offered by ultra-high temperature (UHT, see Chapter 5), means that retorting of most dairy products is much less common for liquid milk products in most countries. There are some exceptions, though. One is small glass jars of infant formula sometimes used in hospitals to feed new-born babies, which have the critical advantage of completely assured safety. Compared to formula supplied in Tetra Pak cartons, the milder heat treatment

*From Farm to Table*. Alan Kelly, Patrick Fox, and Tim Cogan, Oxford University Press.
 DOI: 10.1093/9780197581025.003.0009

(probably UHT) received by the product in the jars is clearly evident from the difference in color, with the retort-sterilized product being markedly browner.

Another product still frequently found in cans in many countries is condensed or evaporated milk, which today is usually used in baking where the "burnt" flavor, due to the severe heat treatment applied, is less of a disadvantage. Because the product is concentrated, lower volumes need to be added in baking or confectionary recipes to provide color and milk solids than if unconcentrated milk was used. In such cases, milk is typically standardized, by having its fat to solids-non-fat contents adjusted to a very precise ratio (typically 9 parts fat to 22 parts solids-non-fat) and then evaporated to the desired level of solids (2 to 3 times that in the starting milk), filled into cans, and heated.

Salts such as phosphate or citrate are usually added to the milk to stabilize it during heat treatment. In their absence, there is a risk that the milk might coagulate in the cans during heating; like many coagulation-related phenomena in dairy products, this is related to the calcium content of the milk, and the added salts prevent these problems through chelation (binding) of the calcium. The level and type of the salts added are chosen based on experience and perhaps small-scale tests and may vary from time to time during the year due to fluctuations in milk salts contents.

The other dairy product still commonly found in cans is sweetened condensed milk (SCM), which is also used in baking, in recipes that require sugar to be added along with milk solids, and so both are provided in one product. In this case, after standardization of the milk, a high level of sugar (sucrose) is added, and the principle of preservation is a little different from that for evaporated milk, because sugar is itself a preservative. If there is a high level of sugar in the environment surrounding a bacterial cell, the force of osmotic pressure acts to try and equalize the sugar (solute) levels inside and outside the cell and draws water out of the cells, dehydrating and killing them. For this reason, the extent of heating required for SCM is not as severe as for evaporated milk, as the sugar provides a powerful additional hurdle against bacterial growth. Sugar has in fact been a popular food ingredient for millennia as much for its preservative properties (especially in products like jam) as for its sweetness.

The presence of sugar has implications for the stability of the product. Hot concentrated milk containing sucrose can hold a very high level of sugars in solution but, as it cools to room temperature, the ability of the water to hold the sugar in solution decreases. In regular evaporated milk, there is plenty of water to dissolve lactose, the only sugar present. However, in sweetened condensed milk, a significant portion of the water is bound by the added sucrose, and there is much less available to dissolve lactose, which is less soluble than sucrose. For this reason, there is a significant risk of lactose forming crystals, which have a characteristic jagged "tomahawk" shape, and are detected as having a gritty or sandy texture on consumption. So, in preparing and cooling SCM, particular consideration is given to controlling the crystallization of lactose to ensure that this defect does not occur. This is typically achieved by "seeding" the product with a small amount of crystalline lactose in the form of small undetectable crystals, which provide a template of the type of structure the rest of the lactose should follow when it is converted from a soluble state into a partially crystalline one.

A related product that is popular in some countries (especially in South America) is dulce de leche (literally, sweet milk), which is made in a broadly similar way to SCM, but is heated much more severely, leading to extensive caramelization and Maillard reactions, both of which give a brown color to this rich, sweet, syrup-like product.

## Sterilization of dairy products

The oldest approach to making milk stable for very long periods of time involved placing it in a can and heating it to a very high temperature, although typically lower than that used for UHT treatment (more like 115–120°C), but for a longer time (perhaps 10–20 minutes rather than seconds).

In such a process, the dairy product is poured into a can made of a suitable material such as tinned steel that conducts heat well, sealed (with a very small headspace between the liquid and the lid, to minimize the amount of air present), and placed in a metal chamber called a retort, which is then itself filled with high-temperature steam at high pressure (which prevents the can contents from boiling) and closed. After a suitable time, the steam is switched off and, as the cans cool, air is gradually introduced to the chamber to balance the pressure, preventing the cans from buckling or deforming due to a sudden pressure change.

A very rapid increase in temperature occurs in the chamber as steam is introduced, followed by a decrease as the steam is turned off. However, this does not reflect the temperature conditions within the can, which is where the bacteria of concern are located, as it takes time for heat to penetrate through the can walls and contents. The key part of the can during heating is called the cold point, i.e., the part of the can that takes longest for the heat to reach. This is also where bacteria receive the least heating and have the greatest chance of survival. If the contents of the can are solid, then the cold point is at the geometric center (furthest from every point on the can walls, base, and lid), because heat travels in straight lines through solid materials (by conduction). If, however, the cans are filled with a fluid, then the cold point is below center, as currents within the fluid will result in hot material continuously rising (the form of heat transfer called convection) and the upper part of the can heats a little faster than the lower part.

It is thus critical for the cold point to receive sufficient heat load (the combination of temperature and time) to inactivate any bacteria, but this should happen as quickly as possible. If it takes a long time for the heat to reach this point, then the remainder of the can's contents will experience far more severe heat treatment, which adds the risk of undesirable consequences, such as color, flavor, and nutritional damage.

How can heat penetration into the can be maximized? Factors such as the can size and shape are important, to give a geometry that minimizes the distance from heat entering the can (through the walls) to the cold point, while the heat transfer can be directly facilitated by rotating or agitating the can during heating, to mix the hot material and help force convection to work more effectively.

The simplest form of sterilization technology is a batch retort, also called an autoclave, which uses high-temperature steam at high pressure to kill all microorganisms present, while preventing the material from boiling (as mentioned earlier). The pressure cookers sometimes found in domestic or restaurant kitchens, and which cook food at a high temperature in a short time, operate under similar principles.

Whereas a domestic pressure cooker or laboratory autoclave might be a small piece of equipment that fits on a table, commercial retorts can be very large, and typically take the shape of a horizontal cylinder, with the door/lid at one end, to facilitate loading and unloading of cans on pallets or trolleys. As one of the easiest ways to enhance the efficiency and evenness of heating the can contents is to mix the contents during heating, in many retorts the cans are rotated during the heating phase.

The biggest disadvantage of retorting is that it is a batch process, involving repetitive cycles of loading, heating, and unloading each batch of cans. Batch processes are very undesirable in the food industry, as they involve a lot of time that is not spent doing the key part of the process (in this case heating) and a lot of labor in loading and unloading the cans.

The ideal process would be continuous, with some cans always entering the retort while some are being sterilized and others are leaving. In the case of canning, the key barrier to the development of such systems is the need for the heating step to be undertaken at high pressure, but having entry and exit points for the system—for cans to enter and leave—would require openings through which steam could escape, preventing suitable pressures from being maintained. This problem was solved by the clever application of a principle in physics called the hydrostatic principle, which states that a tall, thin column of water exerts pressure at the base of the column which is proportional to the height of the column.

A hydrostatic sterilization retort based on this principle comprises two tall narrow columns of water, joined at their base by a chamber filled with steam under pressure. A conveyor belt passes in through one leg, through the chamber, and out the other leg. Cans of food to be sterilized are loaded onto the belt entering the first leg, which descend vertically, gradually increasing in pressure the further down they go (there are no sudden changes in pressure). The cans then enter the steam zone and spend enough time there to be adequately heated, before leaving this zone by ascending through the second tower of water. The pressure of the contents of the cans gradually decreases as they exit the retort, when they are removed from the belt and replaced by new cans ready to be heated. Cans are always being added to the belt, and sterilized cans are always being removed, in a continuous process, suitable for large-scale efficient sterilization.

## Concentration by evaporation

A significant portion of the water in milk can be removed, which will reduce its volume (and weight) and increase the level of solids by a factor of about four (from 10–12% to 40–50% solids). This may be done either to produce a concentrated product like evaporated or sweetened condensed milk or as a prelude to full drying, to make the drying process more efficient by reducing the amount of water that must be removed in the final step.

In concentrating milk, some of the water molecules present in the milk must change from a liquid to vapor, which is more easily done at higher temperatures. However, heating milk to a high temperature risks damaging it through undesirable heat-induced changes to flavor and color, as well as destruction of nutrients and denaturation and changes in the properties of proteins. So it is preferable to evaporate the water at a lower temperature, which is achieved by exploiting the fact that the boiling point of water is reduced when the pressure is reduced. If pressure is the force opposing the outward movement of water from a liquid to a gaseous phase, then reducing that force makes it easier for water to boil (decreases the boiling point) while increasing the force has the opposite effect.

Picture a simple system where milk is flowing vertically downward inside a tube, on the outside of which steam circulates, supplying heat and energy. A vacuum pump reduces pressure, and some of the water in the milk, flowing down the inner wall of the tube, evaporates. Thus, as the milk descends, the interior of the tube becomes filled with water vapor; when it reaches the bottom it is relatively easy to separate the vapor, which rises, from the heavier concentrated milk, which sinks.

Maximizing the surface area over which the reaction occurs is key to this process, which is why the milk flows down the inside of the tube as a thin layer. Increasing the volume of milk being concentrated requires not a bigger tube, but rather a lot of small tubes, which collectively have a much greater surface area. Such a bundle of tubes surrounded by steam is called a calandria.

To make the process even more efficient, a number of calandria (where each is called an effect) can be connected in series, with the milk flowing from one to the next and becoming progressively more concentrated as it does so. In such a case, the vacuum is drawn after the last effect, and the boiling point in each effect after the first one is progressively lower, as the vacuum gets progressively stronger.

In such a system, a major cost is the generation of steam that supplies the energy to drive the evaporation. In the multiple-effect system described, the vapor coming off the milk in each effect is at a higher temperature than the boiling temperature in the next effect, and so steam needs to be supplied only to the first effect, to kick-start the process, with the vapor coming off the milk there being used as the heating medium for evaporation in the second effect and so on. This reduces the steam and energy requirements hugely, and so modern milk evaporation plants typically have multiple effects, with up to seven calandria connected in series.

In this way, a concentrated milk product can be produced that can either be used as a product in its own right (condensed milk is the term often used for milk concentrated by evaporation) or fed directly into a drying process to produce a powder.

## Drying of milk

To convert milk to a powder, it is necessary to rapidly remove water without excessively damaging its sensory and nutritional properties, while yielding a final product that is easily reconstituted with water. The key is to supply a large amount of energy to the milk, which causes the water present to evaporate and be removed in the form of vapor, leaving the dried milk solids behind. The most efficient source of such energy is hot, dry air, and the key to this is to ensure that air and milk mix and react in the most efficient way possible and for the shortest time possible, to avoid undesirable thermal damage to the milk at these high temperatures.

In early processes for the drying of milk (which are still practiced in some small plants), water is removed by roller drying: a thin layer of evaporated milk is poured onto the surface of hot rotating metal cylinders or drums (Figure 9.1). The milk is directly in contact with the source of heat energy (the drum), and the water evaporating from the milk escapes directly into the surrounding atmosphere. The rate of rotation of the cylinders is such that, once sufficient water is removed, a blade scrapes off the powder (in the form of flakes) and the exposed drum surface is covered with fresh milk feed, to continue the drying process.

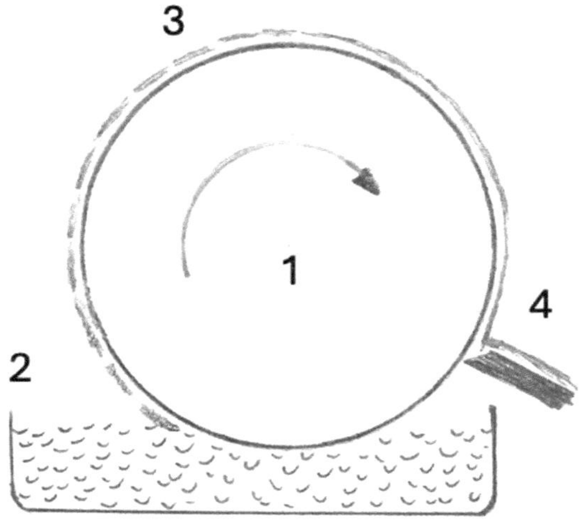


**Figure 9.1** Schematic diagram of operation of a roller-dryer. The rotating metal drum (1) is internally heated by steam; as it passes through the bath of fluid at the bottom of the figure (2), it picks up a thin layer of liquid, which gradually dries due to evaporation of water (3) as the drum rotates, until it is scraped off by the knife at right as powder flakes (4), leaving the drum surface free for application of fresh liquid (image thanks to Anne Cahalane).

This process is efficient but results in a dried product that has undergone significant thermal damage, in terms of color and flavor changes, due to the long contact time between the milk and the extremely hot surface of the roller. The temperatures used were reduced by the development of roller dryers in which the drying system was placed in a vacuum chamber (reducing the boiling point). However, this technology is little used today, as the process damages the fat globules in milk, releasing fat from them (called free fat) which is undesirable for many applications. The main application for roller-dried whole milk powder today is as an ingredient for chocolate, as the free fat gives a desirable, "shiny" appearance to the chocolate (see Chapter 13).

It is preferable, instead of using a hot surface to supply thermal energy, to use hot air. But it is critical to maximize the surface area over which contact of air and milk takes place, by converting the fluid feed of the condensed milk into a physical form that will dry rapidly without suffering excessive heat damage.

In roller drying, milk is spread in a very thin layer, but the heat is being supplied from the hot metal surface and the water being evaporated can escape only in the opposite direction. Can the milk be surrounded on all sides by a source of heat, which would allow the water to escape in every direction?

The principles for drying milk (or indeed any food) have very commonplace analogies in everyday life, such as drying clothes on a clothes line. Under what circumstances do we find "great drying conditions" for that domestic job? Usually, we think of "good drying" as taking place when the air is warm, moving, not too humid, and the clothes are well spread out and hung so as to maximize contact with the air. These are exactly the same principles used when drying food. The product needs maximum exposure to the drying air, and that air needs to be hot (to supply the most energy), dry (to absorb more water), and moving. Movement is important as when water evaporates from a surface, it tends to remain nearby, and could form a cool damp insulating layer, which hinders further drying unless it is removed swiftly and replaced with fresh dry, hot air.

How is contact of milk with air maximized? The shape that provides maximum surface area for any volume is a sphere or droplet, so it is necessary to convert the milk into droplets, which are then sprayed into the dry, hot air. The heat source also acts as the "sink" for the removed water, which means the air has to be very hot and very dry since, the more moist the air is, the less water it can remove from the drying milk.

Making a spray of droplets from a liquid is not very difficult—it is done routinely using garden sprays or nebulizers for medicine or perfume. The process used in drying milk is called atomization, for which there are a number of mechanisms. In one process, the milk is fed through a nozzle where it is constricted and then released through a narrow opening—perhaps forced through with air—and emerges in the form of a cloud of droplets; this is called a pressure nozzle atomizer.

In another design, the milk (pre-concentrated to perhaps 40%–60% solids by evaporation) is fed into a bowl-shaped chamber that is rotating at high speed; the periphery of the bowl has slots in it and the milk, on being driven by centrifugal force through these slots, emerges as a cloud of droplets. This is known as a rotary atomizer.

Technologies that dry milk in the form of a cloud of droplets are collectively called spray-driers. The cloud in such cases emerges into a large chamber which is typically cylindrical at the top, with a conical lower section (i.e., cylindroconical in shape). When the spray enters the upper section, it meets a swirl of very hot dry air (typically at around 200°C). Each droplet then loses moisture, first from the surface and then progressively from deeper within, as it drops through the chamber under the influence of gravity.

The milk never reaches a temperature as high as the surrounding air, as the evaporation of water absorbs energy (the latent heat that is needed to fuel the evaporation), which provides a cooling influence. As the droplets become gradually drier, the surrounding air becomes correspondingly cooler (having lost energy to the droplets) and more moist (as it has absorbed the moisture removed from the milk). Air entering the chamber at 200°C may leave it at 70–90°C; the final temperature reached is a measure of the exact amount of drying achieved and is monitored carefully as an indicator of the efficiency of the process.

Drying chambers used for spray-drying milk (or indeed any other food, e.g. coffee), are frequently very large—several stories high and perhaps 18 meters in diameter—and are probably the largest single pieces of processing equipment used in the food industry. The largest industrial dryers in the world (in plants in Ireland and New Zealand) can produce up to 30 metric tons of powder per hour. Close to the drying tower, the heat and noise are immense during operation. The noise comes from efforts to reduce the danger posed by the accumulation of powder on the inside walls of the dryer, which can get so hot as to generate a dust explosion, that is dreaded by all and is as scary and serious as it sounds. To prevent such explosions (which have led to fatal accidents in the past), modern spray-dryers are fitted with sophisticated fire detection and suppression systems, including cameras and detectors. Another option is to pound the walls of the tower at frequent intervals with massive automated hammers to dislodge adhering powder, which accounts for some of the noise. In addition, as a fail-safe measure in the case of an explosion, the buildings housing drying towers frequently have a weak wall, designed to collapse and release the explosion pressure and resulting fire rather than trapping it inside. These dryers are very serious and complex pieces of equipment to manage safely!

A schematic of a modern spray-dryer is shown in Figure 9.2, and a photograph of a small scale dryer is shown in Figure 9.3.

Air flow within the chamber is carefully managed and directed, and by the time the air has completed its drying mission, it can be channeled out of the chamber quite easily separately from the powder, which falls downward and is removed through an opening at the base of the chamber and on into the storage and packaging operations.

However, the air exiting the chamber contains a small, but significant, amount of powder, typically very small, light particles that were not sufficiently heavy to sink downward with the rest of the powder. The loss of these would reduce the yield of powder from the milk and also present an environmental and pollution issue, if milk powder were to escape from the factory to settle somewhere outside. The exhaust air from the chamber must thus be stripped of this residual powder, and this is done in a number of ways.

One approach is to feed the air through specially shaped chambers, called cyclones, in which the flow pattern is such that the powder particles are thrown to the outside for collection. Cyclones are also used in modern vacuum cleaners, to strip and collect dust from the air being drawn into the system. Another approach is to use filters in the form of bags with pores that are too small to allow the powder particles to pass through (similar to older vacuum cleaners). They

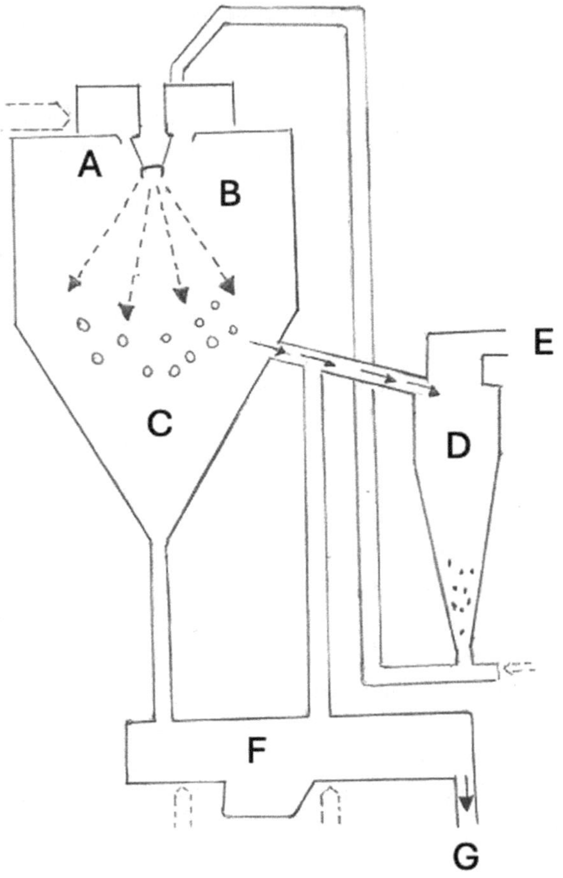


**Figure 9.2** Schematic diagram of operation of a two-stage spray-dryer for the production of milk powder. Evaporated milk enters the chamber at (A) where it is atomized into a fine spray of droplets (B), which are surrounded by hot air and dry to powder particles as they fall (C). The hot air exits the dryer and enters a cyclone (D), where centrifugal flow causes the powder to fall downward while the hot air exits (at E). The powder from the main chamber and cyclone enters a fluidized bed (F), where remaining moisture is removed under gentle conditions, and the now dry powder leaves for packaging (G) (image thanks to Anne Cahalane).

gradually fill with recovered powder, which can be removed later and mixed with the main bulk of powder recovered from the chamber. The final approach is wet scrubbing, where water is sprayed into the air stream to dissolve and thereby remove the fine particles.

**Figure 9.3** A small-scale spray-dryer, with the separation cyclone visible on the left (University College, Cork, Ireland, food processing hall).

Overall, by one or more of these approaches, the air leaving the drying chamber is depleted of entrained powder particles before being vented to the outside of the drying facility. The mixture of powder recovered by one of these methods and the majority of the powder that came directly from the chamber are mixed together and then bagged into large sacks, or otherwise prepared for further use.

Such a powder, however, while better in quality than that produced by roller-drying, is still quite insoluble and not very convenient to use. This is because all the water removal is taking place in the chamber, and requires going from an initial moisture level in the concentrate of 40%–50% down to less than 4% in the powder, all in the space of time it takes to fall through the chamber. So, this drying is harsh and intensive and rather crude in terms of being able to control the properties of the final powder.

To more finely tailor the properties of powder, the key is to make the drying gentler and more gradual. One way to do this is to allow the powder to leave the main chamber at a higher moisture content (maybe around 10%, requiring less intensive heating in the chamber) and fall into a separate drying unit where the remaining moisture is removed.

In this process, called a two-stage spray-dryer, the second stage is called a fluidized bed, which is like a tunnel with a false floor made of perforated metal. Powder enters the tunnel at one end and meets a stream of warm air being blown through its floor. The perforations are too small to allow powder particles to fall through and, at a certain air velocity, the powder falling onto the floor will float as the force of the air overcomes gravity. At a higher velocity, the powder will move along entrained in the stream of warm air (called a fluidized bed of powder) and has been dried to its final target moisture content by the time it reaches the far end. Such a powder will be far more soluble and convenient than that dried in a single stage.

For whole milk powder (WMP), reconstitution can be further helped by spraying the bed of powder with lecithin, a good emulsifier, which helps to overcome the natural repulsion of water by this powder, as will be discussed a little later.

The drying of powder can be made even more gradual and gentle by essentially replacing the conical base of the drying tower with another fluidized bed, onto which the powder falls by gravity for some intermediate drying before exiting to an external fluidized bed. This arrangement is called a three-stage or multi-stage dryer (sometimes called an MSD dryer).

Further control of powder properties can be achieved by precise tailoring of the powder particles produced, rather than each powder particle originating from a single droplet of sprayed milk. More complex structures can be easier to disperse and dissolve in water in their final application and are often comprised of clusters of powder particles agglomerated together in particular arrangements. They are sometimes called grapes or raspberries, names which convey the idea of shapes made of multiple small particles connected together to give a larger entity.

To produce these structures, modifications are made to the atomization zone where the droplets are created, at the top of the drying tower. One method uses two or more nozzles to shoot jets of powder at an angle such that the sprays of droplets they produce collide with each other, which gives the kind of composite structures desired. Another method takes the small light powder particles recovered in the cyclones from the exhaust air and feeds some back into the atomization zone, where they mix with droplets and again dry into agglomerated structures; as these small particles are called fines, this approach is called "fines return."

## Milk powders

As well as a greatly extended shelf-life, milk powders have a range of other benefits, particularly convenience. Storage is simpler, as is transportation, particularly over long distances, of sacks of dry powder that do not require refrigeration and perhaps weigh a tenth of the original material. Dry milk can also be more easily used as a source of milk solids in other formulations, such as bakery recipes, without the large amount of water that otherwise accompanies milk as an ingredient.

For these reasons, the drying of milk has been of industrial and scientific interest for a long time. There are reports that sun-dried milk powder was produced by the Mongols in the time of Genghis Khan in the 13th century, but the first commercial dried products were probably produced in England around the middle of the 19th century. These were made by mixing milk, sugar, and sodium bicarbonate in open pans, but did not yield a satisfactory product.

Roller drying was first used around the start of the 20th century in the United States and United Kingdom, and remained a common means of drying milk for the next half century or so. The poor quality of products resulting from the harsh drying led to the search for more gentle drying methods and the introduction of spray-drying around the middle of the century.

For milk powder to be useful at a later stage, it must be easily converted back to a liquid form; direct consumption of powder is unlikely and so a key property of any powder is its ability to be reconstituted easily. In other words, it must react well with water so that it regains a liquid state, ideally as close as possible to the original milk from which it was dried. Not only that, but it must do it in a manner that is convenient for the intended applications. There is a big difference between a powder that dissolves only after several hours of mixing with hot water and one which can be reconstituted readily by gentle mixing with water at room temperature. So, the speed and ease of reconstitution are important properties.

These are not simple considerations when it comes to milk powders, compared to some other food materials, as many of the components of milk are known for their hydrophobic properties. While lactose is reasonably water-soluble, milk fat is clearly immiscible with water. So, many milk powders are naturally difficult to reconstitute and steps must be taken to overcome this problem.

If a spoon or scoop of powder is added to a beaker of water, it will first sit on the surface of the water, unless some physical force such as stirring or mixing is applied. Ideally, when the particles of the powder come in contact with water (the first step in their dissolution), there should be a large surface area for interaction between water and the powder. Essentially, the powder needs to break apart into the smallest particles possible, each of which can then be surrounded by water; clumping of powder on addition to water is highly undesirable.

One way in which the interaction of air and water can be facilitated is to maximize the presence of air in the powder. Most powders contain significant levels of entrapped air, some of which is found in between the particles (interstitial air) and some of which is within the particles (occluded air). When the powder is placed in water, these air spaces form channels through which water can flow, thereby maximizing its rapid interaction with the milk components and allowing the imbibed water to disperse the powder. The ways in which the content and nature of air present in a powder can be controlled during manufacture relate to the number of stages of drying, and the use of steps like fines return, as discussed above.

So, the water enters the air spaces in the powder, which makes the particles more dense (as water is far denser than air), and so the powder mass starts to sink. It is important to ensure that the mass is agitated during this process, to prevent it from simply settling into an undissolved mass at the bottom of the container. This may be achieved by mechanical stirring or, for products like infant formula, by rapid shaking.

The interaction of powder with water is also more effective the more energetic the water molecules are, in other words the higher their temperature. Milk powders will reconstitute more readily in warm than in cold water.

The first step in reconstituting a powder is called wetting, and the wettability of a powder depends on its tendency to either reject or accept interaction with water. Powders high in hydrophobic materials like fat can be difficult to wet, and so the first steps in their reconstitution can be problematic. The composition of the surface of the powder is a key factor in this, and the presence of hydrophilic compounds like lactose can lead to far higher wettability than if the surface is covered in fat, particularly free, unemulsified fat.

Oil and water do not mix, and thus the fat present actively repels water when powder and water are introduced to each other, making it very hard to dissolve fat-containing powders. To overcome this mutual repulsion, something that makes the fat less hydrophobic must be present. The problem is analogous to the creation of a stable emulsion from oil and water, we help these mutually hostile phases to coexist by having a third species present, an emulsifier, which has both hydrophobic (water-repelling, fat-loving) and hydrophilic (water-loving, fat-repelling) parts to its molecular structure. Such compounds are amphiphilic and can arrange themselves so that the hydrophobic part of the molecule is aligned with the fat phase while the hydrophilic part attracts and interacts with the hydrophilic water molecules. In the case of milk powders containing fat, the amphiphilic compound used to overcome the repulsion of oil for water is a phospholipid called lecithin—a molecule with both a highly charged, and hence water-friendly, phosphate group and a water-repellent lipid part—which is commonly found in many foods, including milk and eggs. To produce so-called lecithinized fat-containing powders, the lecithin is sprayed onto the powder at the fluidized bed stage of drying, and essentially coats the fat-rich regions with water-attracting charged groups, making the powder far more easy to disperse and dissolve when mixed with water.

## The families of dairy powders

The main dairy products dried today are skim milk, whole milk, whey, whey-derived ingredients, and buttermilk. In addition, whole milk powder may sometimes be specified as being instant, which means that it will readily reconstitute when added to water. The complexity of drying whole milk into a usable form, due to the high level of fat present, has been discussed already, but even systems without fat can be challenging to work with.

For example, in skim milk powder and whey powder, the main component present is lactose, which in theory should be an advantage, since it is a sugar and hence more soluble in water than fat or protein. However, lactose in its dry form is a rather challenging material, as it occurs in a number of different molecular forms (see Chapter 2).

In certain forms, lactose tends to react in a less than ideal way with water, absorbing it and forming large interlocking crystals, which results in a problem called caking, where a powder that at one stage is free-flowing and easy to handle solidifies into a solid mass that is very difficult to divide and work with. The key to avoiding caking is to ensure that the correct type of crystals form during the drying process and prevent the uptake of moisture during storage of the powder, by carefully selecting packaging materials and resealing packages after opening.

Instantized powder is an important category of milk powder for many applications—the name refers to how readily these powders return to solution on mixing with water; a powder that disperses and dissolves readily in water is described as instant. Many skim milk powders, particularly those produced in multi-stage driers, are instant, while whole milk powder may be made more instant by lecithinization.

# 10
# The Production of Milk Protein Ingredients and Lactose

As has been discussed in previous chapters, milk is an extremely complex material containing a number of different components, each with very different properties, and the production of many of the basic dairy products involves some type of fractionation of these components. For example, making cheese involves concentrating the fat and casein from the milk, while part of the protein fraction (the whey proteins) and most of the lactose is lost in the whey, and making butter involves concentrating the fat fraction almost exclusively.

So, making dairy products has always involved separating milk into different fractions. But in recent decades, the technologies for doing this and the range of ways in which milk can be fractionated to a wide range of different ingredients have increased greatly.

## The history of whey usage

For as long as cheese has been made, whey has been produced. For most cheese varieties, 90% of the volume of the milk used to make cheese ends up as whey. Whey consists of around 92% water and looks like a very watery, slightly green, fluid, which lacks the richness and complexity of milk. It appears that all the valuable components of the milk have ended up in the cheese, and the residual whey is of much less interest or value. For this reason, for most of the time over which cheese has been made, whey has been seen as a disposal problem rather than an opportunity.

It has thus been common to dispose of whey by feeding it to animals, especially pigs, spreading it on land, for which it is a poor fertilizer, or even just throwing it away. If you look at a map of any country or region containing a number of cheese factories and look at where factories built before the 1970s are located, they are often to be found alongside, lakes, rivers, or the sea. There are even old photographs of lovely beaches onto which large pipes convey large quantities of whey directly out to sea.

*From Farm to Table*. Alan Kelly, Patrick Fox, and Tim Cogan, Oxford University Press.
 DOI: 10.1093/9780197581025.003.0010

Starting in the 1970s, however, there has been a profound reevaluation of the usage and potential of whey. This was driven in particular by environmental concerns: whey has both a high biological oxygen demand and a high chemical oxygen demand—due to the presence of milk nutrients—which results in depletion of vital oxygen in aquatic systems, such as rivers, into which whey might have once been discharged. But things have changed. It used to be said that whey was the low value by-product of making cheese but, today, it is sometimes jokingly said that cheese is the low value by-product of making whey. This reflects the newfound value and hence respect placed on whey.

The first ingredient of interest to recover from whey was the protein. The whey proteins account for around 20% of the protein in milk and appear, at first glance, to be far less useful than casein, which, after all, is responsible for the structure and production of products like cheese and yogurt.

In recent years, as will be discussed later, a wide range of valuable applications for whey proteins have been identified, which have driven the development of methods to recover them from whey. The key principle for this recovery relates to the most exploitable difference between these proteins and the other major components present in whey, lactose and salts. This difference is size, as the protein molecules are thousands of times bigger than the other components.

How can molecules be separated on the basis of size? The most straightforward method is to pass the material through a filter of some sort that retains the larger molecules while allowing the smaller ones to pass through. Picture this in simple domestic applications like sieving powders in the kitchen or filtering coffee grounds. In these cases, the materials being retained and filter pores are very large and even visible, whereas, in the case of whey, all the molecules concerned are soluble in water and of relatively small size, which means that the filters required must be designed with pores of an exceptionally small and controlled size.

The key technology used for this is called ultrafiltration, which was developed initially in the early 1970s. The material used for this filtering is a polymer (plastic) sheet, which appears at first glance to be solid and impermeable, but under an electron microscope can be seen to be actually perforated with extremely fine pores. These pores are typically around 0.01 μm (micrometers) in diameter; as the naked eye can distinguish objects only about one-tenth of a millimeter in size (100 μm), they are far too small to be visible. On this microscopic scale, a whey protein molecule will find it impossible to pass through such a pore, whereas the much smaller salt and lactose molecules will pass through freely. In technical terms, we have separated the whey into two fractions called permeate (which passes through, or permeates, the membrane: the water, salts, and lactose) and retentate (which does not pass through, or is retained: protein, as well as small amounts of lactose and salts).

If whey consisting of 95% water, 4% lactose, 0.7% protein, and 0.3% salts was dried to a powder, that powder would consist largely of lactose and contain only around 10% protein. Adding this to a food or formulation it would also mean adding a lot of lactose, which may well be undesirable. Ultrafiltration, however, reduces the relative contribution of the components other than protein, by removing most of the lactose and salts. If the retentate after a typical ultrafiltration process was dried, the protein level might be enriched to perhaps 35%. A powder made from this is called whey protein concentrate.

In a typical ultrafiltration process, the extent to which protein can be enriched is limited by the retentate becoming more concentrated and viscous; to remove more lactose and salts it is necessary to dilute the retentate with water and then filter the mixture again. This is called diafiltration, which delivers much higher levels of protein, of 60%–80% or even 90%; products containing such levels are called whey protein isolates. Whey proteins today are commonly found as ingredients in many formulated food products, where their ability to gel and thicken are useful. They may even blended with flavors and other ingredients and sold in dry form as protein supplements in gyms or supermarkets, because of their amino acid profile and because they contain essential amino acids (especially leucine, a branched-chain amino acid) which are very good for muscle building.

In a very simple filtration process, a circle of porous paper is folded in the shape of a cone into a funnel, into which the material to be filtered is poured; the filtrate (or permeate) drips through, while solids larger than the pores in the paper are retained. This is called dead-end filtration and is a small-scale, batch process, which ends when the funnel is full of filtered material and must be replaced.

In industrial practice, a continuous process is used, which at its simplest can be pictured as a tube, the walls of which are composed of a suitable filter material, through which the fluid to be filtered is pumped, and which is surrounded by a larger external tube. The liquid feed flows under pressure, which drives any materials that can pass through the membrane into the outer tube to do so and "escape" the pressure; these can then be drawn off. So, as the feed enters the inner tube at one end, it gradually separates over the length of the tube into two streams—that which has flowed outward through the filter (the permeate) and that which has not (the retentate).

In practice, there are a number of filtration system designs, the key being to present a maximum surface area over which the separation can take place. This can be achieved by having the filters arranged in hollow tubes, as fibers, in the form of stacks of plates, or in spiral-wound configurations. The filtration material itself may be a polymer or a ceramic. In all cases, a sheet of the material will look smooth and impenetrable, unless scrutinized carefully under a very powerful microscope, whereupon the tiny pores become visible. In addition, the actual filtration layer made up of these pores typically forms only a very thin layer of the filter, the remainder comprising a more porous support material through which the separated streams can flow easily.

During operation, material that has precipitated or just become lodged near pores builds up on the membrane surfaces and is unable to pass through. This process is referred to as fouling and will gradually obstruct the flow of the permeate and increase the pressure required to force the fluid to flow efficiently. Fouling is typically the limiting consideration for the length of the filtration run—before operations need to stop for cleaning. Much research has been devoted to the goal of minimizing fouling during membrane separation processes.

There are also commercial processes for recovering other proteins from whey, such as lactoferrin, which can be purified by passing whey through special resins under conditions where the lactoferrin bears a different electrical change to the rest of the proteins present, allowing it to be selectively removed. The lactoferrin can then, due its antimicrobial activity, be used in applications such as toothpaste, where it helps prevent microbial growth that otherwise could lead to dental cavities (or caries).

## The value of whey proteins

A major factor in the change in perception of whey as a disposal problem to a valuable resource was the realization that whey proteins are valuable food ingredients. Even though they are present at relatively low concentrations, around 0.7%, membrane separation allows production of materials that are far more enriched in proteins, up to whey protein isolate powders, which might contain around 90% protein in their dry state.

One reason for their popularity is the nutritional value of whey proteins, which is why people seeking to build muscle regard them as a good food supplement—and why giant tubs of dry whey protein-based powders are often sold in gyms and supermarkets. Whereas caseins, on being consumed, form a clot in the stomach due to the presence of acid and enzymes (quite analogous to cheese-making), which is slow to digest, whey proteins are so-called fast proteins, which are digested rapidly to amino acids that contribute to muscle building and repair.

To understand their value and uses, the properties of one of the proteins in particular, β-lactoglobulin, are critical. This protein, absent from human milk, has a very specific structure, which leads to interesting properties (see Chapter 2). In raw milk, it is found in a tightly folded, almost spherical structure, because distant regions of the molecule are bound to each other by disulfide bonds between pairs of cysteine residues.

These strong disulfide bonds are one of the main forces defining the structure of proteins. Proteins without such bonds tend to have simple, relatively linear structures (like the caseins), while the more disulfide bonds a protein contains, the more complex its structure is likely to be. Proteins usually contain an even number of cysteine residues, giving half that number of disulfide bonds holding them in their unique shape; for example, α-lactalbumin, another whey protein, contains eight cysteine residues, giving four disulfide bonds.

Unusually, however, β-lactoglobulin contains five cysteine residues, giving two disulfide bonds and one unpaired cysteine residue, which is quite rare in proteins. In the normal state of the protein in unheated milk (the "natural" state of the protein is referred to as the native state), this unpaired cysteine is found in the core of the molecule, but, when milk is heated above 65°C, the other disulfide bonds break and the cysteine residue is exposed and highly reactive. The protein is at this point irreversibly changed from its original state and is typically referred to as being denatured (since it has lost its native shape).

When the temperature decreases, many proteins that have been denatured due to heat-induced unfolding can refold back into their original (native) shape, and so the process of denaturation is frequently reversible. In the case of β-lactoglobulin, however, cooling does not result in the protein returning to its original shape, and the denaturation in this case is irreversible.

Most significantly, the rearrangement of cysteine residues and formation of new disulfide bonds can take place not just within but between molecules of β-lactoglobulin, so that some of the new bonds formed are actually between cysteine residues on different molecules of β-lactoglobulin. So, in the cooled sample, a new hybrid molecule consisting of two molecules of β-lactoglobulin joined by a strong disulfide bond might appear. This polymerization reaction doesn't have to stop at two though, and long, complex polymers and aggregates of multiple β-lactoglobulin molecules can form.

This all happens because β-lactoglobulin essentially becomes "sticky" when denatured and likes to stick onto other molecules. This can be a monogamous reaction, in which the polymers formed are comprised only of β-lactoglobulin molecules attached together, or polygamous, where β-lactoglobulin has formed a bond with other proteins with cysteine residues. This may include α-lactalbumin, but also κ-casein, the casein in milk that plays a key stabilizing role on the surface of the casein micelles.

This reaction has several implications for dairy products and processing. For example, in milk heated under conditions in which the denaturation of β-lactoglobulin is extensive and a proportion has bound to the casein micelles, these could physically obstruct the access of chymosin to the bond it needs to cleave to initiate coagulation of milk in cheese-making. This results in very slow rennet coagulation and a weak rennet gel. Luckily, under the heating conditions used to pasteurize milk, very little of this takes place but, if slightly more severe heating is applied, then the renneting properties of cheese-milk are more severely affected.

In the case of whey processing, the tendency of β-lactoglobulin to form polymers with itself or other proteins can be a benefit or a problem. A simple solution of whey protein, or pure β-lactoglobulin, with a reasonably high protein content (above 8%), under the right conditions (especially pH and levels and types of salts, such as calcium) will be clear, of low viscosity, and quite like water. However, if a beaker of that solution is heated to 80°C or 90°C for 10 minutes or so, the apparently watery material will transform into a solid gel. Why does this happen? The heat unfolds the β-lactoglobulin molecules, and they interlink to other protein molecules in complex inter-woven chains and networks, to such an extent that the system has solidified. This ability of β-lactoglobulin, or ingredients that contain it, to form solid structures is one of its interesting functional properties, e.g., for making gelled products like desserts.

At low protein levels, or under different conditions, the solution might not gel, but the presence of large protein complexes increases its viscosity, and so whey proteins can act as thickeners. Many soups, sauces, yoghurts, and other products include in their ingredient list a whey-derived product because of this structure-forming ability of β-lactoglobulin.

This phenomenon can also be detrimental, however, as when a beverage containing β-lactoglobulin—e.g., a sports drink intended to take advantage of the muscle-promoting properties of that protein—is heated to kill bacteria and make it safe and stable to consume. In this case, the β-lactoglobulin will denature and aggregate and assemble into large complexes. If these aggregates get too large, the drink might become undesirably viscous, turbid (as the aggregates interfere with the passage of light through the drink), or even prone to settling, as the large protein aggregates fall out of suspension. In these cases, great care in formulating and processing the product is necessary to avoid such defects.

## Recovery of caseins

There are a number of different chemical (e.g., isoelectric point) and physical (e.g., molecular size) principles for the separation of milk proteins, which exploit basic differences in their properties.

The simplest way to fractionate milk proteins is to add acid to milk (typically skim milk, to avoid the unnecessary complexity of fat being present) to adjust the pH to 4.6 over a relatively short period of time. Unlike a yoghurt gel that will form when the pH approaches 4.6 over a couple of hours, rapid acidification results in a chaotic destabilization of the caseins, giving a precipitate of milk protein. This can easily be demonstrated at home by adding vinegar or lemon juice to skimmed milk: small flecks will appear at first, but if the milk is left to stand, and once sufficient acid has been added, a clear lumpy precipitate will settle, which is largely comprised of casein.

In industrial applications, this precipitate can be cut into small pieces, recovered by filtration or centrifugation, washed a couple of times with water to remove entrapped salts, lactose, and whey proteins and increase the purity of the casein, and then dried. As the starting material is rather lumpy and not in a solution, this drying is usually undertaken in specialized dryers (called ring dryers), in which the material rotates at high speeds in a stream of hot air while being mechanically ground to give small dry particles. The resulting product is called acid casein and is used in food applications, such as processed and analogue cheeses, creamers, and some cream liqueurs. It is also used in a range of non-food industry applications, such as paper (to add a glaze to fine quality paper), cosmetics, or as an emulsifier in emulsion paints.

Another easy way to destabilize the casein in milk is by adding rennet; making cheese, after all, is basically a process of separating casein (and fat) from whey proteins. So, in a similar process to that described above for acid casein, the enzyme chymosin (rennet) is added to skim milk, a gel allowed to form, which is then cut, stirred, heated (cooked) to firm it up, recovered by filtration, and washed and dried to give rennet casein, for use as a food ingredient in products such as processed cheese. It can also be polymerized with formaldehyde to give a rigid plastic-like material called galalith, which can be used to make buttons and combs.

Acid and rennet casein have been common dairy commodities for decades and generally contain 80%–90% protein. They differ, however, in their properties, as the process of acidification results in the dissolution of salts from within the casein micelles, which does not happen with renneting. For this reason, rennet casein has a higher ash (mineral) content than acid casein. Both acid and rennet casein are rather insoluble in water, which limits their applications.

To solubilize acid casein, and hence make it more useful, requires the main barrier to its solubility to be overcome, which is its hydrophobicity (water-repelling nature). This can be achieved by adding back compounds that react with the casein to make it highly charged, which attracts water and makes the protein far more able to interact with it and disperse or even dissolve. Alkalis such as sodium or calcium hydroxide serve this function and are added to hot dispersions of acid casein to adjust the pH to close to neutrality (pH 7) and then redrying to give sodium or calcium caseinate, respectively. In the case of calcium caseinate, the metal ion added back is the main one removed by acidification, and the one most central to casein existing in a native micellar form, so in theory, the production of calcium caseinate should restore a micellar structure akin to that found in milk, but in practice, recreating these structures and their original properties exactly is very challenging.

Caseinates are more soluble than acid or rennet casein but are not quite the same as the original form of casein in milk. It is only in recent years that a product has become available that resembles very closely the state of the proteins in milk, and this is called micellar casein concentrate or micellar casein. To produce this requires the gentlest form of separation of casein from whey proteins possible, without chemical or enzymatic modification, using membrane filtration technology (Table 10.1).

In this case, using a particular type of membrane filtration called microfiltration, the pores are of a size that allow casein micelles (in the skim milk used as a feed material) to be retained but whey proteins to pass through easily, producing highly functional casein. This can be used for applications requiring casein with the properties of the micelles in raw milk, such as cheese-making.

A modification of this approach can be used to partially purify β-casein. This is the main casein found in human milk, and so there has been interest in its recovery for application in infant formulae. Microfiltration is used to recover it, but rather than the typical casein recovery processes which involves filtration at

**Table 10.1** Types of membrane separation applied to milk

| Process | Diameter of pores (micrometer) | Milk components passing through | Milk components retained | Applications |
|---|---|---|---|---|
| Reverse osmosis | $10^{-4}$–$10^{-3}$ | Water | All | Concentration of milk or dairy streams |
| Nanofiltration | $10^{-2}$–$10^{-3}$ | Water, salts | Fat, protein, lactose | Demineralization of whey and purification of lactose |
| Ultrafiltration | 0.01–0.1 | Water, salts, lactose | Fat, whey protein | Concentration of protein in whey; concentration of milk for cheese-making |
| Microfiltration | 0.1–1 | Water, salts, lactose, whey protein | Fat globules, casein micelles. Also retains bacterial cells. | Production of extended-shelf life milk; production of casein-rich ingredients |

30–40°C, the process is carried out at a low temperature, using chilled skim milk as the feed.

β-Casein is the most hydrophobic of the caseins and, at low temperatures, hydrophobic interactions decrease greatly in strength, and so the attraction of β-casein for the micelles is reduced at low temperatures, and a certain proportion (up to 20%) dissociates from the micelles (as discussed in Chapter 2). If cold milk is microfiltered, the dissociated β-casein is too small to be retained by the membrane pores and passes through these into the permeate, giving a retentate that is depleted in β-casein and a β-casein-enriched whey protein stream.

To recover the β-casein from the permeate requires the trick that resulted in its separation to be essentially reversed, by encouraging hydrophobic interactions to reform through increasing the temperature. Warming the permeate from a cold microfiltration process causes the β-casein molecules to reassociate and, in the absence of the other caseins, relatively pure β-casein micelles form. Although these are smaller than the casein micelles in milk, they are still too large to pass through the pores in a microfiltration membrane, and so a new cycle of filtration at warmer temperatures separates these from the whey proteins, leading to the production of *β*-casein concentrate.

Thus, a family of protein recovery processes based on membrane filtration has evolved, which originally focused on whey protein recovery from whey using ultrafiltration but now encompass separation of most of the caseins from milk in a relatively pure and valuable state. It should be noted that the processes using microfiltration to recover casein also produce a whey protein stream, and that the streams produced in this way differ from those traditionally produced as a by-product of cheese-making. Specifically, cheese whey is a more complex material which contains a lot more than proteins, lactose, and minerals. Additional components present in cheese whey include:

- caseinomacropeptide, the part of κ-casein cleaved off by chymosin in the act of coagulating the casein micelles;
- starter bacteria cells and active chymosin, which may need to be inactivated before further processing; and
- color such as annatto, sometimes added during cheese manufacture.

The whey produced by microfiltration contains none of these, and so is sometimes referred to as ideal, native, or virgin whey.

As well as microfiltration and ultrafiltration, membrane separation processes with even finer pores are available, such as nanofiltration, in which the pores are so small that only mineral salts and water can pass through. This could be used to reduce the mineral content of whey, for applications such as infant formula for which a low mineral content is desirable.

The finest pores of all are found in membranes used in a process called reverse osmosis, in which the only molecules small enough to permeate are those of water. This process can be used to produce extremely pure water and is used outside the dairy context to desalinate sea water to yield safe potable water.

## Lactose

The component present at the highest levels in milk, after water, is lactose, which also dominates many dairy streams, including most of the solid component of whey. In particular, it is the major component of the resulting permeate streams after removal of the protein to produce whey protein ingredients discussed above.

While lactose is less sweet than sucrose, it has a wide range of applications, though it must first be recovered from the whey. This is relatively easily achieved, by taking advantage of the fact that lactose has limited solubility and, once it becomes insoluble, it forms crystals that can be recovered and purified.

Lactose is thus prepared by crystallization from concentrated whey or whey ultrafiltrate (in which its concentration has been increased to exceed its solubility limits) and the crystals recovered by centrifugation. About 400,000 metric tons of lactose are produced annually, compared to ~100 million metric tons of sucrose.

Crystallizing lactose from whey or permeate requires that the liquid be evaporated to a level of around 60% solids and then cooled. When hot, a large amount of lactose can be held in solution but, when the temperature decreases, so does its solubility, and so when the hot concentrated fluid is cooled, lactose crystals form rapidly. Lactose can form different kinds of crystals, each of which has different properties, so the exact process followed will depend on the specific characteristics and applications intended for the final product. Steps taken to control this can include "seeding" the liquid with a small amount of lactose in the desired crystalline form, which induces the crystallizing lactose to adopt the same form.

The crystals can then be recovered using specialized decanter centrifuges, which remove the solids from the liquid and can wash them free of impurities, followed by drying, grinding to a fine powder, and packaging. Greater degrees of purity can be achieved by mixing a lactose solution with activated carbon, a highly porous product made from charcoal, with a high surface area and propensity to absorb impurities.

Such refined lactose is used in pharmaceutical applications. It can make up the majority of the solids in a tablet, having been mixed with an active ingredient and then compressed into tablet form, to give a solid and manageable pill that can be readily swallowed, or it can be used in an inhaler or nebulizer to "carry" the tiny quantities of active agent being delivered. Its advantages in such applications include its solubility, bland taste, ability to be compressed into tablets, and cost effectiveness, but it may have limitations for individuals with lactose intolerance (although the levels taken in this form do not seem to be a major problem for such people).

Owing to its relatively low sweetness and low solubility, the applications of lactose are much more limited than those of sucrose or glucose. Its principal application is in the production of infant formulae based on cows' milk (human milk contains ~7% lactose compared with ~4.6% in bovine milk (see Chapter 12). Lactose has a number of low-volume, special applications in the food industry, including as a free-flowing or agglomerating agent for powders, enhancing the flavor of foods, and improving the functionality of shortenings and as a diluent for pigments, flavors or enzymes.

Lactose is also an excellent starting material for production of a range of other ingredients, such as alcohol. A number of yeast species can ferment lactose to ethanol and carbon dioxide, and so both recovered lactose and lactose-rich streams such as whey are often used in such processes.

Lactose may also be hydrolyzed by the enzyme β-galactosidase (lactase) to give glucose-galactose syrups, which are sweeter than the starting material and can be consumed by lactose-intolerant people. It can also be used to produce (through fermentation) citric, acetic, and lactic acids, and can also be converted to oligosaccharides (prebiotics, or compounds that promote the growth of beneficial bacteria in the intestines and which may be added to infant formula), lactulose (a disaccharide containing galactose and fructose, which is a prebiotic and a laxative), lactitol (the alcohol form of lactose), and lactobionic acid (a sweet-tasting acid, which is a very rare property, that is used in cosmetics as a skin-softener and moisturizer).

Lactose can also be chemically converted to the compounds ammonium lactate and lactosyl urea for use in animal feed, as they are slowly digested in cows' rumens to yield nitrogen, which is used to produce protein.

# 11
# Ice Cream

Ice cream has been a very popular dessert around the world for a long time. It is reported that Alexander the Great consumed delicacies composed of ice mixed with honey and nectar (in the fourth century BC), there are biblical references to iced drinks, and the Romans favored snow or ice mixed with fruits and juices. Around AD1300, the Venetian explorer Marco Polo is reported to have brought a recipe from the Far East for what is essentially a sherbet. Products similar to ice cream were reported in the royal courts of France and Italy in the 16th and 17th centuries, which later became more widely publicly available. A recipe for a gelato including milk, cream, butter, and eggs was offered at a café in Paris by an Italian, Procopio Cutò, around 1680.

It has been reported[1] that George Washington enjoyed ice cream and served it to guests, but the manufacture and storage of ice cream became widespread only after the invention of related processes such as mechanical refrigeration (by Jacob Perkins in 1834, and later applied on an industrial scale by Carl von Linde in the 1870s), the extraction of sugar from sugar cane and sugar beet, and the increasing use of domestic refrigerators in the early decades of the 20th century.

Today, ice cream is perhaps the dairy product most associated with pure hedonic pleasure and indulgence, both in sensory terms (due to its characteristic creamy texture and sweet taste) and nutritional terms (due to its high content of fat and sugar, which mean it is often being considered as a "treat" that should be consumed in moderation). It is also a very seasonal product, with consumption in general increasing in the summer.

A wide range of ice cream products is available, largely differentiated by their flavor, but also by their mode of consumption: frozen just before consumption and eaten in relatively soft form in a cone or a cup or distributed and stored frozen in tubs or cartons until consumed. In addition, different types of ice cream vary in their composition, with the typical levels of sugar and fat being modified to increase acceptability to more health-conscious consumers, leading to low-fat and low-sugar variants, as well as entirely non-dairy versions (more correctly described as frozen desserts). There are also frozen yoghurt products, for which the principles of manufacture of ice cream and yoghurt are combined to a certain extent, giving products where the dominant sweetness is somewhat offset by acidic and related flavor notes.

*From Farm to Table*. Alan Kelly, Patrick Fox, and Tim Cogan, Oxford University Press.
 DOI: 10.1093/9780197581025.003.0011

Consumers expect certain very specific characteristics of ice cream:

- A smooth texture, with creaminess the dominant sensation and grittiness due to the presence of detectable ice crystals being absent.
- A slow melting rate, such that the product does not run uncontrollably off a cone before eating or collapse too quickly when served as part of a dessert.
- Smooth and rapid melting in the mouth with a pleasurable flavor release.

Achieving these characteristics is, as for many of the dairy products discussed in this book, far more complex than might be expected considering the familiar nature of the product and the fact that it has been made and consumed for so long. The traditional art of making ice cream has been gradually supported by increasing scientific understanding of the phenomena involved, which has helped manufacturers understand how best to control, optimize, and modify the product.

Interestingly, the science behind ice cream is quite different from that of other dairy products. For example, the key disciplines involved in understanding the production and ripening of cheese are chemistry and biology, but in the case of ice cream the key discipline is probably physics, as will be discussed in this chapter.

## The freezing of ice cream

As mentioned already, two of the defining characteristics of ice cream are sweetness—due to the presence of sugar—and suitable melting properties; they are more interlinked than might be expected. For example, a sugar-free ice cream would have a very different texture and melting behavior than a more conventionally formulated one. Why is this the case?

To begin with, consider what happens if milk is placed in a freezer. Does it have a texture like ice cream? No, it simply resembles frozen milk, with a solid texture more like that of water ice than ice cream. The principal difference between frozen milk and ice cream is due to two factors present in ice cream that are absent from milk: one is air and the other is sugar.

Water freezes at 0°C, and to reduce the formation of ice on cold mornings, salt is typically sprinkled on the roads. How does this work? The presence of salt interferes with the ability of water to form ice crystals, and so reduces the freezing point of water. So, if the temperature on a cold morning is −1°C, adding enough salt to reduce the freezing point of water to less than this temperature means less ice forms, and the roads are less slippery. The more salt that is added, the more the freezing point is reduced.

Sugar has a similar effect to that of salt and when added to an ice cream mix makes it more difficult to freeze, by depressing the freezing point. If an ice cream mix containing sugar is placed in a freezer, a temperature lower than 0°C is required to initiate freezing and, when this is reached, some of the water transforms into ice crystals of pure water. The remaining sugar becomes concentrated in a smaller volume of unfrozen water, and this increase in sugar concentration makes it still harder to freeze. While this is taking place, the temperature of the ice cream continues to decrease, and so gradually more ice will form, while the decreasing remaining volume of liquid water becomes more and more concentrated in sugar and progressively harder to freeze.

At the temperature at which ice cream is stored, some of the water will be in the form of ice and the remaining water will be so concentrated with sugar that it cannot be frozen at that temperature, which is above its freezing temperature. In ice cream, this might mean that only 50% of the water is found frozen as ice crystals, which are essentially suspended in a highly concentrated sugar syrup. Ice cream is thus not a solid frozen material but rather a very viscous liquid, within which are suspended ice crystals. A schematic diagram of the structure of ice cream is shown in Figure 11.1.

## The production of ice cream: formulation

Making ice cream depends firstly on the selection of ingredients according to a set formulation or recipe. The first key defining ingredient is the fat, and ice cream is defined in terms of the fat content as full fat (around 14–16% fat), reduced fat (maybe 10–12%), low fat, very low fat, etc. The source of fat in traditional ice cream is milk fat, derived from cream, or perhaps milk, with, in some cases, added butter or butter oil. In lower-price product formulations, some of the fat may be vegetable oil.

The function of the fat in ice cream is very strongly related to the mouthfeel, richness, and indulgent aspects of the product. The fat does not affect the freezing properties of the product directly, but certainly has an indirect effect on the ice formation properties, as the fat globules occupy space within the product matrix and leave less "room" for ice crystals to grow to large sizes. The presence of fat also reduces the amount of water present; in lower-fat products, the lack of this space-filling component can result in a greater risk of iciness or grittiness in the product.

The next characteristic ingredient of ice cream is sugar (typically sucrose). As mentioned earlier, the sugar in ice cream serves dual functions, most obviously in giving the desirable sweet flavor, but less obviously in terms of controlling the freezing point of the water. For this reason, making healthy ice cream is not just a

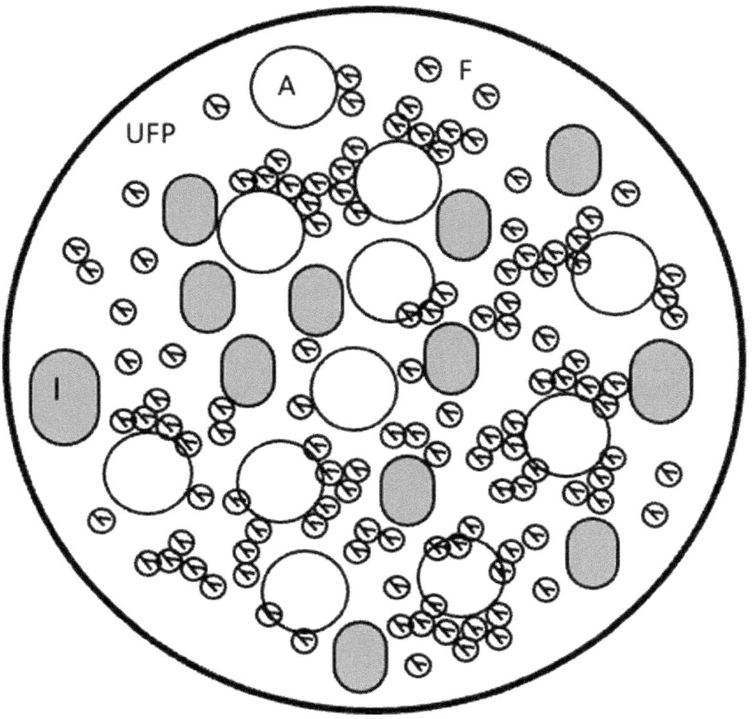


**Figure 11.1** A schematic diagram of the microstructure of ice cream, with (A) air bubbles, (F) semi-crystalline fat globules, and (I) ice crystals all suspended in an unfrozen phase (UFP), which is essentially a highly viscous sugary syrup (image thanks to Professor Doug Goff, University of Guelph).

question of replacing the sugar with artificial sweeteners (like sucralose or aspartame); while the sweetness of ice cream might be matched, the tiny levels of powerful sweeteners would have no impact on the freezing properties of the product.

Of course, milk and cream already contain a sugar, lactose, but lactose is far less sweet than sucrose (less than 20% as sweet for the same level of sugar), and so to produce a sweet product, which freezes in the desired way, additional sugar is usually added. This could be, for example, fructose syrups made from corn (maize) starch in the case of less expensive products, or alternatively products naturally rich in fructose such as honey; fructose is twice as sweet, weight per weight, as sucrose, and is better at depressing the freezing point, as twice as many molecules per weight are added. That is, in a given weight, there are twice as many molecules of fructose as sucrose—because fructose is a monosaccharide, while sucrose is a disaccharide—so less fructose needs be added to a formulation than sucrose.

The next key ingredient is an emulsifier, which may be a mono- or diglyceride or perhaps a natural source of emulsifier such as egg yolk, which is rich in the phospholipid called lecithin. For a molecule to act as an emulsifier, it must be amphiphilic, so that it can sit comfortably at the interface between these two phases, with one foot (molecularly speaking) in each. Mono- and diglycerides achieve this by having one or two hydroxyl ($^{-}OH$) groups, respectively, which are polar (and thus hydrophilic), and either two or one fatty acids, respectively, which form the hydrophobic part of the structure. In the case of phospholipids such as lecithin, the charged part contains a negatively charged hydrophilic phosphate group, while a fatty acid provides the required hydrophobic character.

The role of emulsifiers in ice cream is rather complex. When making the mix, the fat is typically homogenized, as this distributes it evenly throughout the mix and gives a good final structure. However, as discussed in Chapter 5, this means that the globules become coated in protein (casein), which makes them very stable. This is desirable in the case of making a stable homogenized milk product, but in the making of ice cream it is important to be able to whip and incorporate air bubbles into the mix and make a foam; homogenized fat globules are very stable (due to protein being bound to their surfaces) and are not easy to destabilize, as is required in whipping.

For this reason, the fat globules must be partially destabilized to make them whippable, which is the function of the emulsifier. When emulsifiers are added into the homogenized mix, they have a greater affinity for the fat globule than the protein, and so compete for the space at the interface, ultimately displacing the protein at the fat globule surface. These are easier to destabilize mechanically than the protein-coated globules and, when the mix is mechanically whipped in the presence of air during the freezing process, a partially coalesced network forms in which the destabilized globules have agglomerated (to minimize their contact with water) and surround the air bubbles whipped into the mix.

The ice cream mix also typically includes hydrocolloids, usually polysaccharides such as starches or gums, which are referred to as stabilizers. Emulsifiers and stabilizers are often considered closely related ingredients in a formulation, but in practice they have very different functions. While emulsifiers play their key role during the freezing and structure formation process, stabilizers act principally during frozen storage of the ice cream to maintain a desirable structure, as will be discussed later.

The remaining ingredients in the ice cream are critical for consumer acceptance and the desirability of the product, but perhaps have less significance for its manufacture. The presence of most flavors, for example, or even chunks of fruit, biscuit, or chocolate, has little impact on how the ice cream freezes or melts, but has a major impact on who will buy it and allows manufacturers to extend their product ranges.

In fact, this is a nice example of a type of product innovation called line extensions, when a manufacturer has a base product, which they know how to make successfully, and of which they regularly introduce relatively minor variants, typically new flavors. The majority of other ingredients, and the process for making the product, probably remain unchanged ("if it ain't broke, don't fix it"), but one or a small number of things may be removed, altered, or added. Ironically, the most scientifically simple modification can result in the most profound change in consumer perception or desirability. Examples of this, besides ice cream, include new flavors of potato chips (or crisps), bottled water, or soft drinks, where major brands introduce ranges of themed or seasonal flavors.

In terms of developing or modifying a food product, a much more fundamental and generally challenging change is a reformulation, when some much larger changes are made. In the case of ice cream, this would involve altering some of the most fundamental factors influencing the product's characteristics, specifically sugar and fat. The reasons for such modifications relate to manufacturers seeking to address consumer concerns about the unhealthy indulgence of ice cream as a product for which pleasure may be tinged with guilt. However, whereas changing just a flavor typically requires little other change to the product, removing or reducing one of these cornerstone ingredients will have far more challenging implications for the creaminess, melting properties, or mouthfeel of the product.

Another reason for reformulation is to reduce the cost of the product, by replacing expensive ingredients with cheaper alternatives; creating a recipe for ice cream often involves selection from a price list and balancing desired product attributes with considerations around the target market, perceptions, and ultimately profitability. Examples of cost-based reformulations include the avoidance of egg as an emulsifier in cheaper ice creams, the replacement of sucrose by fructose derived from corn starch, and the partial replacement of milk fat by vegetable oils. Fructose has an intense sweet taste that is perceived earlier in the mouth than that of sucrose yet dissipates more rapidly, and so is regarded as a flavor enhancer, while it is also a smaller molecule than sucrose and so is more effective in depressing the freezing point of water. So-called high fructose corn syrups (HFCS) are increasingly unpopular with consumers due to their reported links to obesity, diabetes, and high blood pressure, and because they contain a more rapidly absorbed form of sugar.

Another example of the considerations around the cost of a specific ingredient relates to what is often regarded as the most basic ice cream flavor, vanilla. Natural vanilla is produced by a plant that is a member of the orchid family, and today mainly in plantations in the islands of Madagascar (the source of around 80% of natural vanilla) and Réunion. However, there are significant and increasing problems around security of supply (which can be influenced by everything from weather, climate change, political unrest, and earthquakes, with the Covid-19 pandemic causing particular problems for supply and transportation and leading to a global shortage of natural vanilla), which has resulted in the cost of this natural ingredient increasing. As a result, many vanilla-flavored products today actually contain artificial or synthetic vanilla, in which the key compound vanillin (which has the characteristic "warm," sweet, and creamy flavor) is produced chemically rather than naturally. This is a good example of how consumer demands for "natural" rather than "artificial" on food labels can lead inevitably to increasing cost.

## The production of ice cream: processing

The first step in ice cream production is the selection of the ingredients, to match a formulation designed on the basis of price, availability, target market, and the other considerations mentioned above. The raw materials are a mix of fluid (cream, milk) and powdered ingredients (sugar, stabilizers, etc.), which are blended in the desired proportions, dissolved and mixed, and then subjected to the two fundamental processes used in the early stages of production of most products discussed in this book, which are pasteurization and homogenization.

In the case of ice cream, these serve the key goals to—above all else—ensure safety and extend shelf-life in microbiological and physical terms, preventing both microbiological spoilage and physical separation.

However, the exact process differs due to the specific requirements of ice cream, and the exact nature of the ice cream process, as two different production scenarios can be envisaged.

Scenario A: a producer prepares a formulation, processes it right through to freezing, packaging, and hardening (to be discussed below), and distributes these for sale.

Scenario B: a producer produces a formulation, up the point of preparing a physically and microbiologically stable mix, and then packages this while it is still fluid, for sale and distribution to other locations (such as shops), where the final step of freezing is undertaken just before sale to consumers on cones or in tubs. In this case, retailers purchase the unfrozen mix, typically refrigerated in containers like Tetra Paks, and then freeze this on-site on demand.

In the first scenario, the time gap between production and freezing is less than 24 hours, and so, assuming that freezing stabilizes the product, stability needs to be maintained only for this period, while in the second there could be weeks or even months between mix production and freezing. In the latter case, the stability of the product is far more crucial and, if distribution takes place during the warmest parts of the year (when ice cream demand is likely to be highest), producers may ensure maximum shelf-life and stability by applying a more severe heat treatment, such as ultra high temperature (UHT) processing. In the case of ice cream, the presence of sugar and other strongly flavored ingredients is likely to mask the typical cooked flavor induced by such severe heat treatments, and so they are less likely to put off consumers.

In either scenario, the minimum heat treatment needed is that required to ensure adequate microbiological safety, which means killing all pathogens for mixes which are to be kept refrigerated, and spores in the case of long-life mixes intended for distribution at ambient temperature. The exact time and temperature will be determined by the combination of factors that apply, such as the more sluggish heat transfer characteristics of the high viscosity mix containing a lot of insulating fat and the need to inactivate enzymes (lipases) that could otherwise hydrolyze the milk fat, leading to rancidity, when the globules are damaged during the whipping process. The high levels of sugar also represent an additional antimicrobial hurdle. Typical time-temperature conditions used are around 80 °C for 15–30 seconds, while more intense heating (up to UHT conditions) may be applied for long-life mixes to be stored at ambient temperature.

The mix is then homogenized, typically as part of an integrated heating process as described in Chapter 5, at 40–50°C, before the final step in the heating process.

In terms of what happens next, consider only what was referred to in Scenario A above, where a single manufacturer is responsible for both the preparation and freezing of the mix. In this case, following pasteurization and homogenization, the mix is cooled and held refrigerated for a number of hours before freezing, and often overnight, which is called "ageing" or "ripening" the mix.

During this time, all the milk (or other fat) triglycerides with a melting point between 4°C and 80°C will gradually crystallize into a semi-solid material, which will be perfect for whipping. In addition, this period allows for the molecular rearrangement at the fat globule surfaces to take place, as the emulsifier molecules present displace the milk proteins, taking their place at the interface. Finally, the stabilizer (polysaccharide) molecules unfold and absorb large quantities of water (a process called hydration), ready to function as required during the storage of the final, frozen ice cream.

When sufficient time has passed for all these phenomena to occur, the ice cream mix is frozen. In a domestic kitchen, this can be done in a bowl maintained at a very low temperature, using whisks to incorporate air, while damaging the fat globules, whipping the mix, and freezing a significant proportion of the water present.

In industrial processes, however, a combined whipping and freezing process is performed using a scraped-surface heat-exchanger, which essentially comprises two concentric tubes, in the gap between which flows a freezing medium (Figure 11.2). The mix flows through the inner tube and comes in contact with the extremely cold wall separating it from the freezing medium, which induces rapid formation of ice crystals. To prevent the rapid formation of an immobile sheet of ice, the center of the tube contains a rotating metal shaft, from which protrude a set of blades that continuously scrape the ice off the walls, distributing it through the mix as small crystals and transporting fresh mix to the wall, to be frozen in turn. The action of the blades also incorporates, in the form of small bubbles, the only other ingredient going into the inner barrel besides the mix, which is air.

At the end of this process, the ice cream mix exits the scraped-surface freezer at a temperature around −5°C, as a mix of air bubbles and ice crystals in a viscous sugary syrup. The product is then in a fluid and flexible state and can for example form a self-sustaining (albeit time-limited, depending on the outside temperature) conical shape, in the form of a classic ice cream cone. In retail outlets, the ice cream freezer is a scraped-surface heat-exchanger, and the product going

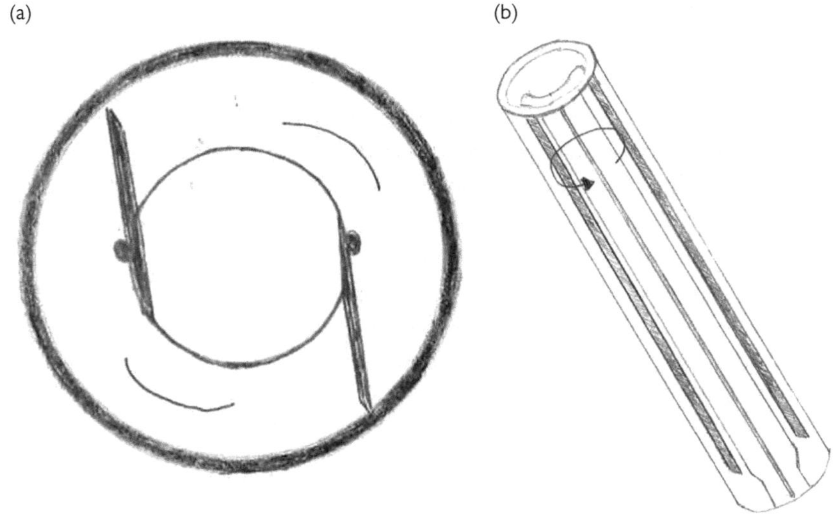


**Figure 11.2** A scraped-surface heat exchanger used for freezing ice cream (in a factory or in retail premises), shown (a) end-on or (b) side-on. The ice cream mix flows through the inner tube, and ice forms on the inside wall due to the presence in the outer tube of an extremely cold refrigerant. The rotating blades attached to the central shaft continuously scrape the inner wall, removing the ice as tiny crystals and distributing them through the ice cream mix (image thanks to Anne Cahalane).

onto cones or into tubs is the semi-frozen mix ready for eating. This is called soft-serve ice cream.

If the objective is not immediate consumption, the product could be filled into tubs ready for hardening by lowering the temperature to around −18°C, at which temperature it will be stored and distributed. The softness of the product at the point of packaging provides an opportunity for the addition of other ingredients that would not perhaps have been suited to processing steps like homogenization, pasteurization, and continuous freezing without either breaking down or being altered in their flavor or character (due to heat-induced changes). These might include pieces of cookie or biscuit, chocolate, or fruit, which can be added at the soft semi-frozen stage, once care is taken to ensure that this does not re-contaminate the mix with undesirable microorganisms.

The temperature of the mix is then reduced to the final temperature, and a significant additional proportion of the water present now freezes. A key property of the final product is the size of ice crystals present, as the human tongue is very sensitive to particles more than 20 μm (0.02 mm) in diameter, above which an undesirable gritty or sandy texture is perceived—whether the particles in question are sugar, ice, or other crystals. The more slowly something containing water is frozen, the larger will be the ice crystals produced, and so when freezing any food product, it is important to freeze it as quickly as possible, to give the smallest crystals possible, which are the least perceptible, but also the least damaging to food structure. In the case of ice cream, this means that the soft-frozen but packaged mix will be blasted with ultra-cold, fast-moving air, in tunnels, on belts, or in spiral systems, to reduce the temperature as quickly as possible.

This should give a smooth frozen product of good consistency ready for distribution, sale, and home storage. But how is that consistency maintained during storage, which can last several months? That is the function of another ingredient present, the stabilizer.

## Maintaining quality during frozen storage

Formulating ice cream correctly and freezing it properly is only the first step in delivering a high-quality product to the consumer. It is critical that this quality is maintained through frozen distribution and home storage in the consumer's freezer. What are the main challenges in this regard? The key issue is actually to control the size and growth of ice crystals during this period of storage.

To understand what the problem might be, consider a simple domestic scenario. If ice cubes for drinks are made in an ice cube tray and then placed in a plastic bag for storage, the cubes in the bag gradually start to stick to each other over time and eventually begin to fuse into larger frozen structures. Why does this happen? The problem is the temperature in the freezer is not constant. In particular, the temperature fluctuates every time the freezer door is opened, which allows in some warm air. This causes a very small amount of melting on the surface of the ice cubes, and when the temperature decreases again this liquid water on the surface of the ice cubes refreezes, but is likely to join neighboring ice cubes together. This is why, over time, ice cubes gradually fuse together, because they are being bridged by ice formed from water that melted when the temperature increased (even briefly).

In the case of ice cream, the principles are similar, but the ice cubes in this case are ice crystals. A key property determining the eating sensation of the ice cream is the size of the ice crystals formed during the freezing process, which can cause grittiness if too large. So, to maintain consumer acceptability, it is critical to make sure that the ice crystals are below this size, and the greatest threat to this is the growth of crystals due to thawing and refreezing, even to a tiny extent, during frozen storage.

Just like the ice cubes in the earlier example, every time the temperature of the ice cream increases even by a couple of degrees, some of the ice thaws and, when the temperature decreases again, this tends to refreeze in an uncontrolled manner, which leads to a gradual and inexorable increase in crystal size in the ice cream. If this is not controlled and the growth limited as much as possible, the ice cream will become unacceptably gritty. Anyone who has found ice cream left at the very bottom of a freezer and forgotten about for a long period of time and then tastes it will understand exactly what this is like.

Putting off this gritty icy fate for as long as possible, by controlling the mobility of water when it thaws, is key to making high-quality ice cream, and this is the role of the polysaccharide stabilizers, which bind and immobilize large quantities of water. Stabilizers in ice cream are typically gums or starches that form a network in the ice cream around the ice crystals, so that any water that thaws due to temperature fluctuations is kept close to the original ice crystal rather than migrating. Then, when this water refreezes, ice crystals should return to their original size; the risk of them fusing and increasing in size uncontrollably will be greatly reduced, and the shelf-life of the ice cream thereby greatly extended.

So, overall, the production of a high-quality ice cream depends largely on the selection of the right ingredients—all of which have a function in controlling the flavor, texture, or frozen shelf-life of the final product—and applying the correct processes to yield the desired structures. Once again, the apparently simple pleasures of a much-loved dairy product are derived from a complex set of scientific principles—in this case very much determined by the science of physics, rather than the highly complex biology and biochemistry that determine the characteristics of a product like cheese.

The basic ice cream can then be modified in a myriad of ways, as mentioned above, by selecting different ingredients and inclusions (such as flavors, colors, caramel, biscuit, or chocolate pieces). Ice cream can also be solidified inside molds to give ice creams on sticks, with the sticks being inserted during the freezing process when the texture is soft enough to allow it but not too solid; the molds can be warmed slightly to allow the ice cream to be removed, and it can be coated with a layer of iced fruit or chocolate, for example. Frozen yoghurt can be made by freezing a fermented yoghurt mix to which has been added some ice cream ingredients such as stabilizers, or which has been blended with a traditional ice cream mix. Sherbets and sorbets contain low levels of dairy ingredients (or none in the case of sorbets), and are typically based on high levels of sugar and acid flavorings.

As the texture of ice cream is very dependent on the structure of the ice crystals present, and their size is greatly influenced by the rate at which the mix is frozen, different textures can be produced by freezing at very low temperatures, and very fine textured ice creams have been produced by freezing "dots," or small spheres, of ice cream in liquid nitrogen at an extremely low temperature (−196°C), although storage and distribution of products at such low temperatures is beyond most common supply chains.

Overall, ice cream, a very familiar and popular product, as shown in Figure 11.3, is a remarkably complex physical entity—semi-frozen liquid, emulsion, and foam—the desirable properties of which are attained by careful application of science-based principles of formulation and processing.

**Figure 11.3** The appearance of frozen ice cream as it starts to melt (image thanks to David Waldron).

# 12
# Human Milk and Infant Formula

The only complete food created by nature for human consumption is breast milk. Not only must this milk provide complete nutrition at a time when no other nutrient sources are being consumed, but it must fulfil this function at a time when the body is at its most vulnerable, changing and developing rapidly, and most dependent on milk as a sole source of sustenance.

The optimal food for newborn infants is unquestionably human milk but, for those mothers who are unable to breastfeed their infants or choose not to do so for various reasons, alternatives are required. This has led to the emergence of infant formulae, perhaps the most complex category of dairy product in terms of formulation and processing.

Early substitute foods for babies included evaporated milk, which was cheap and sterile and played a role in reducing infant mortality, and roller-dried whole milk. The first milk-based formula produced specifically for infant nutrition was developed by the German scientist Justus von Leibig, who patented the product "Perfect Infant Food" also known as "A Soluble Food for Babies" in 1867. This consisted of milk, wheat malt flour, and potassium bicarbonate, and while it is difficult to see today how it represented a suitable alternative to human milk, it became popular. A modified milk-based infant formula called "Synthetic Milk Adjusted (SMA)," introduced in the United States around 1915, was much closer to human milk in its levels of fat, sugar, and protein, and the SMA brand remains in use today.

The principle of production of any infant formula is based on taking the milk of a different species and modifying it to make it more suitable for consumption by human babies. Many individuals regularly consume milk (or dairy products made from milk) of different species such as cows, sheep, goats, or buffalo every day, but adults do this as part of a diverse diet. But a baby is dependent on human milk as its sole source of nutrition, at least for the first few months of life. In addition, during these months the baby's requirements are incredibly precise and demanding, as it undergoes a period of growth and development far more intense than at any other stage of life. Any milk-based alternative to human milk must, therefore, resemble human milk as closely as possible so as to deliver exactly what the vulnerable infant needs.

How closely does the milk of different species resemble human milk?

*From Farm to Table*. Alan Kelly, Patrick Fox, and Tim Cogan, Oxford University Press.
 DOI: 10.1093/9780197581025.003.0012

The milk of all mammals contains broadly the same basic components: water (usually the dominant component), protein (typically casein and whey proteins), fat (in the form of emulsified globules), sugar (lactose), and minerals. However, there are large differences between species, as discussed in Chapter 1, not just in the relative levels of these (e.g. levels of protein or minerals) but in the exact composition and properties of these families of components (e.g. specific fatty acids, exact caseins, and whey proteins). The principal domesticated dairy animals—cattle, sheep, goats, and buffalo—produce milk of broadly similar composition. Of these, sheep milk has the highest fat content, but the composition of goat milk fat is quite different in the specific fatty acids present from that of cows or sheep, giving products made from such milk a stronger taste. In particular, there are large and relevant differences between species in the relative proportions of caseins and whey proteins, and the exact proteins that comprise each fraction.

The question might reasonably be asked as to whether, despite its widespread availability, bovine milk is actually the best basis for making infant formula, given the differences in composition between bovine and human milk, and the great complexity required in transforming ingredients largely derived from bovine milk into a form that more closely (but unquestionably still imperfectly) matches human milk. For example, the mammal whose milk most closely resembles that of humans is probably horse (mare) milk. The dominance of bovine milk as a base for making infant formula is probably due to the ready availability of a wide range of ingredients and technologies for their modification. Nonetheless, this situation is beginning to change; in some countries, such as The Netherlands, infant formula based on goat milk is now commercially available.

Before considering how infant formula is produced, it is important to first consider the composition and properties of the gold standard it seeks to emulate—human milk—in some detail.

## The components of human milk

As mentioned above, bovine milk without modification is not ideally suited for human infants. The reason for this can easily be seen from a comparison of the composition of bovine and human milk (Table 1.1). Bovine milk has much higher levels of protein and minerals and a lower level of lactose. These differences have significant consequences for the nutrition and developmental benefit the milk provides to the infant; they can even be detrimental, e.g., in the case of the high levels of minerals present. High levels of calcium can inhibit iron absorption and an infant receiving more minerals than required must excrete the excess, which places stress on their kidneys and may result in dehydration due to the high level of urine that must be produced.

A closer comparison of their composition reveals more subtle and significant differences between the two milks. For example, it is not just the level of protein but the types of protein that differ: bovine milk protein comprises 75% casein and 25% whey proteins, but in human milk the proportions are typically 60% whey proteins and 40% casein. The composition of each group of proteins also differs, with human milk being dominated by β-casein and α-lactalbumin and lacking β-lactoglobulin.

In addition, the nature of the proteins is very different in human and bovine milk. As discussed in Chapter 2, the caseins in bovine milk are found in elaborate complexes called casein micelles, whose exact structure has been the subject of intense study, due to their central role in critical phenomena such as the formation of gels during the manufacture of cheese and yoghurt. Because human milk is obviously not processed in similar ways, there has been far less intense study of human casein micelles, but microscopic analysis has shown them to be smaller, less dense, and more diffuse than their bovine counterparts. These differences are probably related to the fact that human micelles are dominated by β-casein and have a much lower content of calcium phosphate than bovine micelles.

As well as the whey protein fraction of human milk being dominated by α-lactalbumin, several other whey proteins, such as the iron-binding protein lactoferrin, are also present at higher levels in human than in bovine milk. This is significant because lactoferrin, and products released from it by enzymatic digestion, have antimicrobial properties due to their ability to bind the iron that bacteria need for growth and survival. Other whey proteins present, despite their apparently low levels, have significant biological and protective roles, such as lysozyme—which kills bacteria by attacking their cell walls—and immunoglobulins—which have major roles in mediating allergic responses.

There are also a number of minor nitrogen-containing compounds (called non-protein nitrogen) that are, although present at very low levels, very significant for infant nutrition. These include taurine and choline: taurine is an amino acid (not found in proteins) with roles in fat absorption, especially in low birth weight (e.g. premature) babies, while choline is important for brain development.

Human milk contains more lactose than bovine milk (typically around 7%), and its carbohydrate complement is more complex, as it includes a diverse range (reportedly over 100 different complex sugars called oligosaccharides are present) that have very specific functions in infant development, including protection against pathogens and assisting neural development. Their role in protecting against pathogens is thought to relate to their acting as "decoys"—they bind to either the microorganisms themselves or to the site where the bacteria attach to cells in the body, thus disrupting the critical "docking" mechanism by which bacteria or viruses can enter cells to cause harm. Such protective effects may explain why breastfed infants suffer far fewer digestive complaints than formula-fed ones, and has led to significant interest in adding oligosaccharides, produced by fermentation or chemical synthesis, to infant formulae. Oligosaccharides are also thought to have prebiotic properties, stimulating the growth of beneficial bacteria such as bifidobacteria in the infant gut, by acting as their main energy source.

The oligosaccharides in human milk are produced by adding sugar residues on to lactose. The simplest human milk oligosaccharides (HMOs) are derived simply from lactose molecules extended with glucose or galactose residues like a string of beads—adding galactose molecules gives galacto-oligosaccharides (GOS). Other HMOs, however, are more complex, and can include fucose, sugars with nitrogenous groups (glucosamines), or acidic sugars such as sialic acid. The list of common HMOs in milk includes compounds with names like 2'-fucosyllactose, lacto-N-neotetraose, and lacto-N-fucopentaose.

After lactose and HMOs, the most abundant component of human milk is fat; as in bovine milk, fat in human milk is found in emulsified globules surrounded by a barrier, the milk fat globule membrane (MFGM), which in human milk shows unusual tendril-like structures in human milk but not in bovine globules. Left to stand, human milk will readily separate into cream and a skimmed milk phase. The lower skim phase is generally far more translucent than bovine skim milk, though, because the main cause of the white color (through the scattering of light), the casein micelles, are much smaller in human milk and present at a much lower level, and thereby less effective at scattering light.

Human milk fat contains many of the same fatty acids as bovine milk, but it has no butyric acid and higher levels of unsaturated fatty acids. In addition, the specific structure of the triglycerides (TGs)—the arrangement of fatty acids—in human milk differs from those in bovine milk or vegetable fats and oils. The properties of TGs, including their digestibility, depend not only on which fatty acids are attached to the glycerol backbone, but at which positions. In general, human milk TGs have saturated fatty acids (like palmitic acid, $C_{16:0}$) at the *sn*-1 and *sn*-3 positions and an unsaturated fatty acid (like oleic acid, $C_{18:1}$) in the middle (*sn*-2) position. This specific structure of human milk TGs is sometimes referred to as OPO (due to the alternating positions of oleic and palmitic acids); the synthesis of OPO fatty acids for inclusion in infant formulae has been an active area of research in the ongoing humanization of infant formulae from bovine milk.

Other critical components of human milk for infant growth, despite the very low levels present, are vitamins, both water-soluble (B, C) and fat-soluble (A, D, E, and K). In most cases, the levels of vitamins present in human milk are more than sufficient for the baby's needs (as might be expected, given that breast milk has evolved to be the sole source of infant nutrition at its most vulnerable stage of life). But in some cases (depending on factors like maternal diet) the levels of vitamin D and K can be low.

Minerals in human milk are also nutritionally highly significant and, although their levels are lower than in bovine milk, they have in many cases very significant roles in the growing infant's body, such as that of calcium in building healthy bones and teeth. Minerals in milk are generally divided into macroelements and trace minerals (or microminerals), depending on the relative levels present. Calcium, sodium, potassium, chloride, magnesium, and phosphorus are macroelements; iron, copper, zinc, manganese, iodine, molybdenum, chromium, fluoride, cobalt, and selenium are trace minerals. The roles of each element have been studied, and a wide range of biological functions identified, including muscle contraction (magnesium), nerve impulses (potassium), glucose metabolism (copper), wound healing (zinc), and protecting cells from oxidative damage (selenium), as well as being a component of the thyroid hormones, thyroxine and triiodothyronine (iodine).

## Factors affecting the composition of human milk

The composition of human milk is not fixed and constant, but is rather variable and depends on a wide range of factors, relating both to the mother and the needs of the infant. Some key factors include maternal diet, gestational age of the baby at birth, time after birth, and maternal illness (such as mastitis).

The milk produced immediately after birth is colostrum, and for the first 48 hours after birth is the most enriched and biologically active form of milk, as the mother's body seeks to provide the vulnerable neonate with the best start in life. After the colostrum period (which lasts 2–4 days), milk is sometimes referred to as transitional milk, and after a couple of weeks becomes mature milk.

The component that is probably most affected by time after birth is protein content, which is very high in colostrum and transitional milk and declines to around 1% in mature milk. Not only does the absolute level of protein change, but so also does the exact profile of proteins present. Colostrum has higher levels of whey proteins, as much as 80%–90% of the total protein at first, decreasing to 50%–60% a few weeks after birth; casein accounts for the remaining protein. Whey proteins meet critical biological needs for the newborn infant, especially immunoglobulins, which are supplied to the infant by the mother *in utero*. Caseins have more of a nutritional function; they are a source of amino acids and calcium and other associated minerals. From this perspective, the priority for the very young infant is protection from disease and establishing a strong immune system outside the mother's body, but, as the baby becomes stronger, this kind of direct maternal support is less critical.

The stage at which a mother gives birth is also significant for the composition of human milk. While a typical full-term pregnancy is around 9 months, babies, for a number of reasons, can sometimes be born much earlier. Premature infants, who are tiny and have very low birth weights, may be born or delivered in case of medical emergency following as little as 25 weeks of gestation.

Since so many of the precursors of milk components come from the diet, it is not surprising that maternal diet has a significant impact on human milk composition, and particularly the fat fraction; there is a strong relationship between dietary fat intake and milk fat composition (more so than fat level). Vitamin levels in milk are also linked to maternal vitamin intake. Examples of the link between diet and milk composition include higher levels of specific fatty acids (omega-3 fatty acids) directly attributable to high levels of fish consumption and a higher level of vitamin C in milk of mothers taking supplements of this vitamin.

Mothers may also suffer bacterial infections of the mammary gland (mastitis) that will, if symptoms persist, be treated with antibiotics. In any case, the composition of milk is altered while the infection is ongoing, as the defense systems of the mammary gland activate to eliminate the threat, and typical changes include higher levels of blood-derived whey proteins and enzymes.

Overall, the study of human milk continues to illustrate the enormous complexity of this precious biological material. A guiding principle in the interpretation of the findings of such work is that everything has evolved to serve one function, which is to support the optimal development of the baby. No other natural material we have come to regard as food has a similar evolutionary origin, and research on human milk has consistently identified significant dissimilarities to bovine milk.

The challenge in producing infant formula is then how to take one material (bovine milk) and modify it so as to offer a reasonable substitute for another (human milk), when at first glance the materials are biologically similar but on closer examination differ in numerous ways, each of which must *mean something important*, in terms of the needs of the most vulnerable consumers, babies.

## Processing and stabilizing human milk

Another active area of current research on human milk concerns its preservation, which is of interest due to factors such as the need for mothers to store their milk, expressed at home, when their babies stay in hospital longer than usual, e.g., in the case of premature infants. Mothers with excess milk also sometimes donate samples of their milk to milk banks from where, once deemed safe for use, they are supplied to mothers whose own supply is insufficient for their infants.

In these cases, human milk becomes a product that needs to be held and maintained in a safe and stable form for periods lasting from hours to perhaps days or even longer. While milk from a healthy mother should be free of bacteria, once it is expressed (e.g. using a breast pump) it becomes susceptible to contamination from a number of sources, including the mother's skin.

A simple way to preserve human milk, particularly in the home, is to freeze it, which stops undesirable chemical and enzymatic reactions and prevents microbial growth. It is important that thawing is undertaken safely (e.g. overnight in a fridge) although occasionally problems such as coagulation or separation of milk on thawing can occur, due to the fact that during freezing, ice crystal formation can damage the fat globules, puncturing them and releasing some free fat. In addition, as water is "removed" in the form of ice, minerals become more concentrated, which can sometimes destabilize the proteins present in the milk.

Another common form of preservation, particularly in milk banks, is pasteurization, to eliminate any pathogenic bacteria that might have entered the milk. Obviously, the volumes of milk available are much smaller than for commercial pasteurization of bovine milk, as discussed in Chapter 5, and so simpler, smaller-scale heating solutions are required. Human milk pasteurization often involves heating milk in plastic or glass bottles in a water bath for around 30 minutes at 63–65°C, which is in fact equivalent to the "low-temperature long-time pasteurization" historically practiced for bovine milk, and still used occasionally in small scale processing facilities.

Such a treatment will render the milk safe and increase its usable life, but has been reported to impair biological value, through denaturation of proteins such as immunoglobulins. To avoid such changes, a number of alternative processing strategies have been suggested. This has paralleled, and drawn from, research into the use of technologies such as high-pressure (HP) processing (discussed in Chapter 15) to produce milk that is as close as possible to raw milk in all ways except microbiological risk, which is also the objective for the processing of human milk. It has been shown that human milk can indeed be preserved by HP treatment (which also has the advantage of being suitable for small-scale sample treatment) without losing the benefits attributed to heat-labile proteins. It seems likely that such technologies will become more commonplace in human milk banks in the future.

Alternative strategies to give a more long-term stable form that can later be reconstituted in sterile water have also been studied, such as freeze-drying human milk: the milk is frozen and then warmed very gently under a vacuum, so ice sublimes to water vapor without going through the water stage and can be removed. It will be interesting to see if technologies used for gentle preservation of bovine milk that we have discussed throughout this book (such as membrane filtration) may yet be applied to this most valuable and fragile of resources.

## Ingredients for infant formula

To produce a product that resembles human milk as closely as possible, the production of infant formula today involves two different, but interdependent, production principles: formulation and processing, each of which will be discussed in turn.

In terms of formulation, the simplest way to make a very basic infant formula could be to dilute bovine milk with water (to reduce the protein and mineral components) and then increase the content of carbohydrate by adding sucrose or perhaps honey. Indeed, this is one of the oldest ways to make a human milk substitute, and was apparently practiced during the 19th and early 20th centuries (although breastfeeding was then probably far more widespread than it is today, and so there was less need for such products). Early formulae also apparently commonly contained ingredients such as wheat flour and starches as sources of carbohydrate. Today, while some countries have very high breastfeeding levels, e.g., Croatia at >90% and Scandinavian countries around 60%, rates in the United States and United Kingdom are lower, around 30%–35%.

In the early 20th century, more advanced approaches in processing and understanding how to make dairy products in general, and better scientific data on the differences between the two types of milk, led to significant improvements in formulation. In a further step in the evolution of infant formula, skim milk could be mixed with a whey protein source (to increase the relative proportions of whey proteins and casein), and lactose added, along with a mix of vegetable and milk fats and oils, to reflect the fat profile of human milk.

The mineral profile is, as mentioned earlier, critical, and so the ingredients used are usually demineralized, using processes such as membrane filtration, ion-exchange, or electrodialysis—a form of filtration in which the filter membranes are charged and so can accept or reject ions and selectively absorb or deplete them from a protein-rich fluid. The precise levels of minerals required by the infant are then added back, as well as vitamins and other essential nutrients.

The ingredient list for infant formula is hence very long, but just a few (the protein sources, fat, and lactose) account for most of the product, with a large number of other ingredients being added at very low, but nutritionally significant, levels.

The protein fraction of infant formula is, obviously, critical, to provide the neonate with essential amino acids but, as mentioned already, the proportions of the major proteins in bovine milk are significantly different from those in human milk. For this reason, using conventional sources of bovine milk protein, such as skim milk or milk protein concentrates or isolates, would result in an excessive level of casein and insufficient whey proteins, which could affect both digestibility and coagulation in the infant stomach. Thus, at a minimum, the proportion of whey proteins must be increased, by adding whey protein concentrates or isolates, to end up with whey proteins representing around 60% of the total protein; the casein source will likely be fluid milk, milk powder, skimmed milk powder, or milk protein fractions such as caseins. The final total protein level should also be much lower than in bovine milk, as the level of protein in human milk is only around a third of that in bovine milk. However, infant formulae usually have a slightly higher protein level (1.3%–1.5%) than human milk (1.0%–1.1%) so as to ensure the correct levels of certain amino acids (such as tryptophan and cysteine) given the differences in amino acid composition between human and bovine proteins. Specific individual proteins that can be obtained in purified form, such as the antimicrobial protein lactoferrin, are also increasingly added to formulae; in recent years, ever more tailored protein profiles have been developed in formulae that more closely resemble human milk.

Due to their very different levels and complexity in human and bovine milk, there has been much interest in recent years in adding oligosaccharides to formula. The most feasible way of doing this is through chemical synthesis, and indeed GOS can be produced from lactose using the enzyme β-galactosidase under carefully controlled conditions. A different type of oligosaccharide, fructo-oligosaccharides (FOS), are sometimes added to infant formulae as a prebiotic; while not digested by the infant, they are very good at stimulating the growth of beneficial bacteria, such as bifidobacteria. FOS comprise linear chains of molecules of fructose, which is not found naturally in bovine milk but is common in fruit and vegetables. Going a step further, some infant formulae today include added probiotic cultures (as discussed in Chapter 5) to directly supply such beneficial bacteria to babies; the formulae are not fermented, but the bacteria are included in a protected state in the dry formula.

The lipid fraction of infant formulae usually represents around 3.5% of the liquid product. As with other components, bovine milk fat does not adequately simulate the fat profile of human milk, specifically in terms of the relative proportions of saturated and unsaturated fatty acids and their exact structure, such as the OPO TGs mentioned earlier. In addition, there are differences in the levels of fatty acids that, although present at very low levels, have critical biological functions, such as ensuring the correct neurological development of the infant. Thus, the lipid fraction of infant formula is carefully designed by adding non-bovine fats, such as vegetable oils, to be blended with the bovine-derived fat, which may come from whole milk, cream, or anhydrous milk fat. Vegetable oils typically added in the past included soy and palm oil, due to the latter's high content of palmitic acid. Palm oil is, however, less popular today, due to health concerns about its impact on calcium absorption as well as the environmental impact of its production.

Other ingredients might include low levels of biologically critical lipids, such as omega-3 and omega-6 polyunsaturated fatty acids (PUFAs), as well as long-chain PUFAs such as arachidonic ($C_{20:4}$) and docosahexaenoic ($C_{22:6}$) acids, which are thought to be important in development of visual acuity and cognitive function.

## Processing of infant formula

The key considerations in processing reflect the need to blend all the selected ingredients and raw materials into a physically and microbiologically stable product that is safe for consumption (given the vulnerable state and immature immune systems of babies). The exact processes involved may differ based on the physical form of the final product, which may be liquid for direct consumption or powdered for reconstitution in sterile (boiled and cooled) water before consumption.

In the case of liquid ready-to-feed (RTF) formulae, safety is typically ensured by the application of sufficient heat treatment to render the product sterile. This may be ensured through either UHT treatment (as discussed in Chapter 5) followed by aseptic packaging into containers such as Tetra-Paks or first packaging into glass bottles or jars followed by sterilization in retorts.

In the production of powdered formulae, the thermal processing applied is generally milder, since undesirable changes (e.g., microbial growth or enzymatic activity) cannot take place in a dry product.

Powdered formulae can be prepared in two ways: dry-blending of dry ingredients in the appropriate proportions or making up a liquid (a "wet") mix which is then spray-dried to give a powdered formula.

Dry-blending is relatively straightforward once all the ingredients are available in a dry form. But, as no additional heat treatment is applied, there is a high reliance on extremely high levels of hygiene being adhered to by the manufacturers of the individual ingredients. Also, various dry materials may differ in properties such as particle sizes and densities, and may physically separate in containers during storage, which could lead to difficulties in later reconstitution of the product.

Producing a formula by wet-mixing is a more common approach, as it allows manufacturers complete control over all stages of the process, and thereby the quality and properties of the finished formula. The manufacturer selects all the materials to be used, of which there are two types: aqueous phase—protein ingredients, lactose, water-soluble vitamins, and minerals—and oil phase—vegetable oils and fat-soluble vitamins. Infant formulae are emulsions, with many ingredients being soluble in one phase (oil or water) but not in the other; they must be prepared separately and then mixed in the appropriate proportions (probably close to 96 parts of the aqueous phase to 4 of the oil). There are two key aspects of preparing a stable emulsion: addition of a suitable emulsifier to keep the two phases in a stable form and application of physical force to disperse the oil phase within the aqueous one, such as homogenization.

The emulsified mix is then subjected to evaporation to increase the solids content prior to drying, which would at one time have been done on roller dryers, but today almost universally involves very large multi-stage spray-driers (see Chapter 9).

The dried powder is then packaged, frequently in metal cans which, before sealing, are flushed with nitrogen gas to remove oxygen and protect oxygen-sensitive components from oxidation during storage (particularly biologically significant unsaturated fatty acids present, such as docosahexaenoic acid).

If, rather than drying, a more severe heat treatment is applied to the wet mix, then an RTD formula will be obtained. In this case, after pre-heating the mix and homogenization, some water-soluble vitamins may be added, before final heat treatment either in the package (in a retort) or by UHT followed by packaging in a Tetra Pak or plastic bottle.

The difference between the types of heat treatment applied are actually quite obvious if three types of infant formula are visually compared: that reconstituted from powder, that in a convenient Tetra Pak, and that in a glass jar. The three clearly differ in color, being progressively darker (more brown) in the order mentioned, as the more intense heat treatment leads to a greater extent of Maillard reactions.

The processing of infant formula also has to take into account the need to deliver the required nutrients in the formula at the levels needed by the infant, and the fact that heat can damage nutrients such as vitamins. To compensate for this, the levels added to an infant formula might be higher than those required, so that, after some losses due to thermal processing during manufacture, the remaining level is the required one; the higher levels added are referred to as "overages."

In recent years, new infant formulae have appeared, including capsules that can be used in coffee-machine-like systems to fill sterile bottles and soluble tablets that dissolve readily in warm water. In addition, formulae based on non-bovine milk have been developed (especially brands based on goat milk, which can have less allergenicity problems than bovine-based formulae for some infants), as well as a number of dairy-free ones, typically based on soy-derived ingredients, but these will not be considered further here.

## Tailoring infant formula

Not only must an infant formula be tailored to the very precise needs of infants, but these needs also change during the first year of life, and the needs of 6 - 12 month old infant are different to those of a 1–6 month old or a newborn baby. For example, in human milk the proportions of the different caseins and whey proteins change during the first year postpartum, and so formulae may be designed to have progressively more casein and less whey proteins when they are intended for use by older infants.

Some infants may suffer allergies to certain components of bovine milk, and formulae in that case must be modified to reflect this. One key example of this is that the bovine whey protein β-lactoglobulin is not found in human milk, and may elicit an allergic reaction in some individuals. The most common way to deal with this is to add proteolytic enzymes to hydrolyze the proteins to small peptides that do not elicit the immune response. Such formulae are referred to as hypoallergenic and may be partially or extensively hydrolyzed. The enzymes added to hydrolyze β-lactoglobulin are not specific and will break down the other proteins also, leading to formulae where most of the protein-related material is comprised of peptides and amino acids, which are easier for infants to digest and absorb. Such formulae are sometimes referred to as "comfort formulae" and may be designed to have a higher viscosity; they are also intended to help reduce the incidence of conditions such as constipation and the digestive upset sometimes called colic.

Another potential negative reaction to infant formula based on bovine ingredients concerns infants who cannot digest lactose, as they lack the enzyme β-galactosidase, which may be linked to colic. In this case, the solution is quite straightforward, as the lactose can be hydrolyzed using β-galactosidase. Parents can do this themselves, as most pharmacies sell preparations containing this enzyme, which can be added to the reconstituted formula and allowed to act for a short period before feeding the baby.

In many areas of science, major advances in understanding come about through the use of increasingly powerful analytical tools. Our understanding of complex mixtures such as milk, has been greatly enhanced by the deployment in the last 20 years or so of so-called "omics" approaches, in which exquisitely sensitive instruments are used to separate complex mixtures of hundreds or thousands of compounds, which are then identified by measuring their molecular mass with very high degrees of precision. The mass of a particular molecule is so specific to an individual compound that automated searches of vast databases of masses of known molecules can allow identification of all the different species present in the complex starting or resulting mixture.

In the case of human milk, proteomic analysis has shown that while most (more than 95%) of the protein present in the milk of any species is comprised of a mix of ten or fewer major proteins (the caseins and major whey proteins), the remaining 5% or less is a mixture of hundreds of other proteins, many of which were unknown before the advent of these analytical methods. Individually these may be present at extremely low levels, but proteins—such as enzymes or those that might regulate growth or other biological processes—can be extremely powerful biological agents, which do not need to be present at very high levels to make their presence highly significant.

Similarly, lipidomic methods give far more detail than ever before on the types of fatty acids present, while metabolomic techniques have profiled very small nutrient molecules and oligosaccharides. Collectively, these methods have greatly expanded what is known about milk (whether human, bovine, or of other species) while providing new indications of what infant formulae should ideally contain to more closely align with human milk.

Much research interest in recent years has thus focused on further humanizing infant formula, by adjusting its composition to even more closely reflect that of human milk. Proteomic studies have revealed the impact on the protein fraction of milk of factors like time after birth, maternal diet, and whether the infant was born at term or prematurely. Better understanding of how to fractionate specific bovine milk proteins has also driven developments in humanizing infant formula, to enrich the proteins of interest and deplete those present at lower levels than in human milk. For example, human milk contains proportionately more β-casein and no β-lactoglobulin compared to bovine milk, so methods to increase or reduce these proteins, respectively, in ingredients for infant formula are obviously of interest.

As well as the proportion present, human and bovine milk proteins differ subtly in their exact chemical properties, specifically in terms of the number of phosphate groups present in proteins (at the amino acid serine). Bovine β-casein is highly phosphorylated, whereas human β-casein occurs in a range of levels of phosphorylation, from none to many. That human casein micelles are far smaller and less dense than bovine ones is perhaps due to differences like this.

To take account of the absence of β-lactoglobulin in human milk, a range of technologies have been developed to make whey protein ingredients that are proportionately enriched in α-lactalbumin and depleted in β-lactoglobulin, sometimes by exploiting differences in their relative solubilities at different pH values and in the presence of different mineral salts.

In terms of the lipid fraction of infant formulae, much research has been devoted to humanizing the lipid profile, and methods have been developed to chemically modify vegetable oils to match the most common triglyceride profile found in human milk fat, which is the so-called OPO structure. Such structured lipids may improve fat digestibility and calcium absorption and reduce constipation.

In conclusion, infant formula today is possibly the most complex dairy-based product. Rather than producing products directly from bovine milk by biological or technological transformation, it must be essentially deconstructed into its constituent parts and then selectively reassembled into a new formulation, with multiple non-dairy ingredients also being added to ensure the best possible match to the composition of human milk. Intense scientific research continues to further narrow the gap between formula and human milk, and production of formula is today undertaken under some of the most stringently hygienic conditions (far more akin to a pharmaceutical plant than a cheese factory), with the ultimate goal of producing the best possible product for babies, who are dependent on that product as their sole source of nutrition and sustenance at the most critical stage of their lives.

# 13
# Milk Chocolate

The cacao plant (*Theobroma cacao*) is indigenous to the Amazon region but has spread throughout Central America. Until the 1800s, cocoa was consumed as a liquid drink, called *chocoltl* by the Maya and Aztecs; it was considered to be an aphrodisiac, and the Aztec emperor, Montezuma, is reported to have been a big consumer. There is archaeological evidence that drinking chocolate preparations were consumed in Mexico as early as 1900BC and that the mucilage surrounding the nibs of the cocoa beans was used in the production of alcoholic drinks. Cristopher Columbus brought cocoa beans to Spain as a curiosity and Hernán Cortés introduced a cocoa-based drink to Spain in AD1520. Over the next 200 years, drinking chocolate spread throughout Europe and became quite fashionable.

The cocoa bean contains about 50% fat and masses or lumps of free fat are formed when cocoa powder is added to hot water, making the drink look unpleasant. This problem was solved in 1828 by a Dutch man, Coenraad van Houten, who developed a press to remove about half of the fat from the beans. A solid eating chocolate was then produced by adding the excess fat to a mixture of cocoa powder and sugar. In 1876 in Switzerland, Daniel Peters added milk solids to a mixture of ground cocoa bean nibs to produce milk chocolate. Roller-dried, whole milk powder was used as a source of milk for chocolate because it was cheaper than fresh milk and contains a high level of free (non-globular) fat, which means that less fat needs to be added during the processing of chocolate to coat the particles and improve flavor.

Milk chocolate is the most popular type of chocolate today, representing about 50% of all chocolate. Milk fat influences the flavor, texture, and quality of chocolate, contributing to a smooth texture and glossy appearance. The milk fat may be added as whole milk powder (~26% fat, preferably roller-dried, because the high free fat content of that product, undesirable in many applications, is valued for the sheen it gives to the chocolate), anhydrous milk fat (99.8% fat), cream powder (40%–70% fat), milk crumb, or as chocolate crumb (as discussed below). Milk crumb is a type of milk powder that is severely heated to 70–100°C for 3–8 hours during manufacture; it contains <2% moisture and undergoes extensive Maillard reaction and carmelization, giving it intense flavor and color. Yoghurt powder is sometimes used in the manufacture of white chocolate, which

*From Farm to Table*. Alan Kelly, Patrick Fox, and Tim Cogan, Oxford University Press.
 DOI: 10.1093/9780197581025.003.0013

contains sugar but no cocoa powder; the acidic taste of yoghurt masks the sweet taste of the sugar.

Today, the cacao tree is grown across the equatorial regions, especially in Brazil, Ivory Coast, Ghana, Nigeria, Indonesia, and Malaysia. Tiny flowers grow on the trunk and branches of the cacao tree and develop into pods, each of which contains 30–45 cocoa beans. The pods are broken by cutting them with a machete or cracking them with a wooden club, and the beans are removed and placed in piles or in large boxes where fermentation occurs. This takes about 6 days and causes an increase in temperature, which kills the embryo of the beans. The beans contain two cotyledons (referred to as nibs). These are covered with mucilage consisting of polysaccharides (including pectin), which are metabolized to ethanol during fermentation by yeast; in turn, the ethanol is converted to acetic and lactic acids by lactic acid bacteria and *Bacillus* spp., which affect enzymatic reactions and contribute to flavor development in the chocolate. Some proteolysis also occurs during fermentation, with the formation of free amino acids that react with polyphenols, producing brown pigments and flavor compounds. Cocoa is rich in polyphenols, especially catechins and anthocyanins, which cause astringency and bitterness.

The fermented beans are dried, either by the heat of the sun (in low humidity environments) or by hot air to about 6%–8% moisture. The dried beans may be milled to cocoa powder at the producing farm or packed in jute sacks and shipped to the chocolate manufacturers for milling. Fermentation and drying are referred to as curing.

The processing of the dried cocoa beans into cocoa powder involves cleaning, roasting at 110–140°C for 45–60 minutes to kill bacteria and develop color and flavor (mainly through Maillard browning), cracking the shell around the beans and recovering the cocoa nibs, and then grinding and milling the nibs to give a product called cocoa liquor. This is then pressed to release about 50% of the fat; the expressed fat is called cocoa butter, and the press cake is called cocoa powder. Soya bean lecithin is usually added (at 0.2%–0.6%) to reduce the viscosity (by emulsion formation) during subsequent chocolate production.

## Chocolate manufacture

Solid chocolate was developed in 1847 by Englishman Joseph Fry, by mixing cocoa powder, sucrose, and cocoa butter, with or without milk powder; this is still the mixture used to produce chocolate. In 1919, Fry's company merged with Cadbury to form Fry-Cadbury, now known as Cadbury. Cadbury evolved from a company founded in Birmingham in 1824 by John Cadbury; it sold tea, coffee, and drinking chocolate.

The mixture is ground using a machine known as a conch, developed by Rodolphe Lindt in 1879, which improved the quality of solid chocolate. The original conch consisted of a granite roller and a granite trough, with a small gap in between. The roller rotates and moves backward and forward for several days, during which coarse agglomerates in the mix (cocoa powder and crystals of sucrose) are reduced in size (to less than 30 μm, to give a smooth consistency) and some acidic molecules are lost to the air. Conching has two equally important aims:

1. To develop flavor.
2. To convert the crumbly paste, flake, or powder that is obtained during refining to a flowable liquid that can be poured into a mold.

During conching, the moisture content is reduced from ~1.6% to <1% due to an increase in temperature, and some unwanted flavors are eliminated. The viscosity decreases due to the loss of moisture and the presence of the soya bean lecithin.

The final step involves tempering at 27–29°C to control the crystallization of cocoa butter. If this is not properly controlled, crystallization will occur on the surface of the chocolate during storage, causing a white finish, referred to as "bloom." Bloom is caused by storage above 24°C or by fluctuating storage temperature. The ideal storage temperature for chocolate is 17°C at a relative humidity of 50%. Milk fat reduces the risk of bloom and usually prevents it.

Three basic types of chocolate (or couverture) are produced:

- Dark chocolate: produced from cocoa powder, sugar, and cocoa butter, containing up to 70% cocoa solids.
- Milk chocolate: cocoa powder, sugar, cocoa butter, and milk solids, containing about 50% cocoa solids. The proportion of milk solids specified for milk chocolate varies, depending on the region, e.g., in the EU, minima of 14% milk solids and 3.5% milk fat are prescribed, while, in the United States, 12% milk solids and 3.4% milk fat are prescribed.
- White chocolate: contains 35% cocoa butter, 14% milk solids, and 2.5% milk fat.

Vanilla is sometimes added as a flavor agent to these products. The typical composition of various chocolates are shown in Table 13.1.

The chocolate industry is valued at about $128 billion per annum; about 50% of which is in Europe and 20% in the United States.

**Table 13.1** The composition of typical chocolate (%)

| Ingredient | Dark chocolate | Milk chocolate | White chocolate |
|---|---|---|---|
| Cocoa liquor | 40 | 12 | – |
| Cocoa butter | 12 | 19 | 23 |
| Milk powder | – | 20 | 30 |
| Sugar | 47.4 | 48.5 | 46.5 |
| Lecithin | 0.5 | 0.5 | 0.5 |

## Chocolate crumb

Chocolate crumb is an ingredient widely used in the manufacture of milk chocolate. Developed in the 1870s by Daniel Peters at Vevey, Switzerland, as a base for milk chocolate, it gives a caramel flavor and revolutionized the production of milk chocolate. In 1905, Cadbury started the production of milk chocolate in England using chocolate crumb, while Milton Hershey started its production in the United States. The advantages of using chocolate crumb as the base for milk chocolate are that it has good storage stability, it can be produced when milk is readily available and cheap, it is usually produced by dairy companies and sold on to chocolate manufacturers, it simplifies the process of chocolate manufacture and reduces the length of the conching period necessary. The flavor of milk chocolate is developed mainly during the manufacture of chocolate crumb, due largely to the Maillard reaction.

Chocolate crumb is produced by adding sugar to fresh or concentrated milk or adding water to a mixture of sugar and milk powder. In either case, this liquid is dried to 80%–90% total solids, cocoa liquor is added, and the mixture is dried in shallow trays under vacuum at 75–100°C for 4–8 hours. The finished crumb contains <1% moisture and is reduced to a powder that can be quickly and economically incorporated into the chocolate manufacturing process. During heating, the Maillard reaction occurs and gives the crumb a brown color and a unique caramelized flavor. The composition of full-fat chocolate crumb is 59% sucrose, 25% milk solids, 15% cocoa solids, and 1.0% water.

This flavor is characteristic of British dairy milk chocolate, which is traditionally made from crumb. The level of caramelized flavor in the milk chocolate can be varied according to the amount of crumb added, which is typically between 15% and 60%. Standard crumb contains 6.8% cocoa liquor, but a 13% cocoa liquor crumb is also available that has a stronger cocoa flavor. Chocolate crumb is also used to deliver a unique caramelized chocolate flavor to chocolate paste/

spread, ice cream, sauces, cakes and cake mixes, drinks, and dairy desserts as a better replacement than cocoa powder. Using crumb allows the manufacturing process to proceed more quickly and economically than by incorporating the separate ingredients, milk powder, sugar, and cocoa liquor. Crumb that contains no cocoa liquor is called milk crumb, which also has a very special and characteristic flavor profile, due to the crumb process.

## Chocolate-containing foods

Chocolate is a component of a wide variety of other foods, including:

- drinking chocolate, available as a powder containing cocoa powder and sugar, prepared by dispersing the powder in boiling water and adding whole, low-fat, or shimmed milk;
- chocolate milk, prepared by adding cocoa powder and sugar to milk of variable fat content. It may be prepared domestically for immediate consumption but a commercial version, heat-treated and packaged, is available in some countries, e.g., the United States;
- alcoholic drinks, e.g., crème de cacao and many cream liquors;
- flour confectionary, produced in various formats, probably the best known being Black Forest gateaux;
- biscuits, with a wide range available commercially, in which chocolate is present throughout, as a surface coating, or as chips;
- sugar confectionary, of which several products are available, usually with a surface coating of chocolate;
- praline (some forms);
- solid ice cream, usually with layers of chocolate within the ice cream block;
- whipped ice cream, in a cone in which a stick of chocolate has been inserted.

Overall, chocolate has been for centuries, and still remains, a popular application of dairy products. It represents a significant application for specific types of milk powders produced in the dairy industry today.

# 14
# Packaging of Dairy Products

Packaging is often overlooked in terms of its importance to food, being seen as simply a bag, box, or carton in which food is transported and stored. However, food packaging today is a highly sophisticated business and plays a key role in maintaining the safety and quality of dairy products.

For food in general, packaging serves a multitude of functions beyond being a container for the product. As well as the obvious protection from contamination, physical damage, or tampering, packaging can protect against a range of factors that could destabilize food, from light (which promotes fat oxidation) to oxygen (which also promotes oxidation, as well as potentially facilitating microbial growth).

Packaging is, above all else, a barrier, and the selection of packaging materials (such as glass, card or cardboard, metals such as aluminum, plastics, of which there are many types, or some combination of these) is based on the properties of the food being packaged and the risks against which the package must mitigate. Each packaging material has particular advantages and disadvantages, and often the packaging selected is a composite of several materials, to synergistically achieve the desired properties. For example, many milk cartons are laminates and careful dissection would reveal a composite structure including layers of cardboard, aluminum foil, and plastic, each of which plays a specific role.

In addition, a food product may be surrounded by several types of packaging. For example, soft cheese may be wrapped in foil, then placed in a card outer box on which is printed the cheese-maker's logo, nutritional information, etc.

To consumers, packaging may appear to be a very simple aspect of a food. But many aspects of it, though essentially unseen, are vital contributors to the food's properties. Fruit or salad, for example, is often bought packaged in a bag, surrounded with what may appear to be air but is actually a mixture of the same gases present in air (namely nitrogen, oxygen, and carbon dioxide) but in different proportions, in what is called modified atmosphere packaging (MAP). Such gases are selected to provide specific functions, e.g., preventing microbial growth while maintaining attributes consumers associate with freshness, such as color and texture. The overall functions and considerations in selecting a packaging material for any food product are summarized in Table 14.1.

*From Farm to Table*. Alan Kelly, Patrick Fox, and Tim Cogan, Oxford University Press.
 DOI: 10.1093/9780197581025.003.0014

**Table 14.1** Functions of food packaging and materials that serve these functions

| Function | Reason | Suitable packaging |
|---|---|---|
| Exclusion of light | Prevent oxidation | Metal foil, metal cans, card |
| Physical strength | Protect from breakage, leakage | Glass, cans |
| Exclusion of moisture | Prevent microbial growth, chemical activity | Plastics, glass, cans, laminated foil |
| Control of gases present | Inhibit microbial growth or oxidation | Plastics |
| Exclusion of gases | Inhibit microbial growth or oxidation | Plastics, glass, metal |
| Printability | Bear marketing, nutritional, consumer, producer information | Paper, card |

In addition, packaging serves a marketing and information function, providing an opportunity to list the ingredients included along with any related concerns such as the presence of allergens, nutritional details, storage and handling conditions, and product and manufacturer names and logos. Other necessary properties include convenience of handling, transportation, and use and, unfortunately increasingly important today, the ability to inform consumers if the product therein has been interfered with (i.e., tamper-evident packaging, such as a seal that must be physically broken on first opening a package).

## Properties of principal packaging materials

Probably the oldest material used for dairy product packaging is glass, which has been formed into containers for hundreds of years and is used today for bottles and jars. Glass has a "premium" quality and offers strong physical and chemical protection (being impermeable to moisture, gases, and most other things), while displaying the product therein. However, it has the significant disadvantages of being heavy and breakable, and, if completely transparent, not offering a barrier to the frequently destabilizing influence of light, which can cause oxidation and off-flavor development in fat-containing products like dairy products. Glass can be colored by adding substances like chromium oxide (dark green), cobalt oxide (blue), or iron oxide (brown).

Today, for many applications, glass has largely been replaced by plastic as a packaging material. In reality, plastic is not a single material but a family of materials with a common chemical structure, comprised largely of long chains of hydrocarbons, which are carbon chains bonded to hydrogen atoms. Minor structural modifications to these long chains make the difference between materials that are transparent or opaque, rigid or soft, or more suitable for forming thin films or solid cartons.

The first plastic developed was polyethylene (PE) in the 1930s. Today, a wide range of materials are produced, which share properties of being chemically inert (not reacting with the food), resistant to microbial growth, and capable of being formed into a range of shapes and formats. The properties of the main plastics used in dairy packaging are summarized in Table 14.2. Plastics used in dairy packaging can be rigid containers (jugs, tubs, cartons), films, parts of composite packaging materials (e.g., as a coating on foil or card), labels, lids, and closures and bands used to show if a package has been tampered with.

The main rigid plastics used in the dairy include PE, which comes in a range of variants differing in density (such as low-, medium-, and high-density forms, known as LDPE, MDPE, and HDPE, respectively), all of which are suitable for forming films but differ in their flexibility and toughness. PE is one of the simplest polymers, made of repeating chains of ethylene molecules, each of which has two carbon atoms and four hydrogen atoms. A PE molecule, then, is a long chain of carbon atoms with linked hydrogens; the length of the chain and structural features such as branching will determine the specific type of PE.

**Table 14.2** Properties of main plastics used in dairy packaging

| Plastic | Properties | Applications |
|---|---|---|
| Polyethylene | Different density variants available (low, medium, and high, LDPE, MDPE, and HDPE, respectively), high flexibility | Forming films; milk bottles and yoghurt cartons (HDPE) and caps and labels (LDPE); component of laminated Tetra Paks; cheese packaging |
| Polypropylene | Translucent, strong barrier | Films and rigid packages (e.g., yoghurt); bottle caps |
| Polystyrene | High and porous | Insulated packages |
| Polyethylene terephthalate | Transparent and light; tough | Opaque milk bottles |
| Polyvinyl chloride | Very flexible, shrinks when heated | Cheese wrapping |

Polyethylene terephthalate (PET) is commonly used in soft drink bottles, as it is transparent, light, and tough. Polypropylene (PP) is translucent but forms strong barriers and is stable at high temperatures; it can be formed into both films and rigid packages and is very heat-resistant, and so can be used in microwaveable packages.

Polyvinyl chloride (PVC) is a very flexible plastic that can be shaped into a wide range of packaging styles, from thin films to rigid transparent containers. Importantly, it shrinks when heated and so can be used to form very tight seals around food products. Polystyrene (PS) containers are very light and porous and are used for hot products (e.g., coffee cups), as they contain a lot of trapped air, which acts as an insulator.

A key property of plastic materials used in dairy packaging, such as PP and PET, is that they are thermoplastics, which means that they are soft, pliable, and moldable at high temperatures and become rigid on cooling, which allows them to be shaped and molded into the required form or structure. One way in which packages are formed is called blow-molding: a solid piece of soft molten plastic is extruded into a mold and sterile air is blown in through a hole drilled through the plastic, causing it to expand and adopt the shape of the mold.

Aluminum foil, first used around 1913, has the advantages of being a light, flexible, attractive packaging material, which is a good barrier to gas and light, but it is susceptible to tearing or puncturing, and so in practice is always strengthened by coating it with a protective layer of plastic. Stronger metal containers used widely for food include aluminum cans, which are typically filled, sealed with a metal lid, and then heated. This resistance to high temperatures makes them useful for sterilized food and dairy products; the food sterilized in the sealed container remains stable for long periods before opening, in some cases up to year or even more. Cans may also be made of tin-plated steel. Metal cans have the advantage over glass of being stronger, less breakable, and providing a tough light-, moisture-, and gas-proof barrier, and have been used for almost two centuries for long-life food products. Cans are made either from two or three pieces of metal, in the latter case one piece being rolled into the cylindrical wall, with two other round disks forming the base and lid. Closure of the lid is critical and is achieved by rollers that bend the overlapping lid around the lip of the can to give a tight sealed closure.

Card, also known as cardboard or paperboard, is another commonly used food packaging material. It is rigid, so can be formed into three-dimensional shapes like boxes and cartons, printable, and a good light barrier. However, it can be weakened by absorbing water or oil and must usually be protected by layers of plastic between it and the food being packaged.

## Packaging of the principal dairy products

### Fluid milk

Glass has been used for packaging milk since the 1880s; in some countries, the standard package for pasteurized milk remains glass bottles with a foil lid (aluminum foil coated with LDPE). In the past, bottled milk was unhomogenized, and the clear package allowed the cream layer, which separated during refrigerated storage, to be seen. As well as the risk (particularly in the days of doorstep milk delivery) of the shiny lids attracting birds, glass has many other disadvantages, such as weight in transportation and breakability. They also lack a barrier to light, and when milk was supplied in glass bottles and held outdoors after delivery, exposure to direct sunlight warmed the milk, encouraged microbial growth, and negatively impacted its shelf-life.

Light penetrates deeper into low-fat or skim milk than full-fat milk. It causes oxidation that can result in flavor changes and the development of a "flat" taste, which reduces the perception of fresh flavor. It also causes nutritional changes through vitamin degradation: light destroys riboflavin (vitamin B2) and catalyzes the loss of ascorbic acid (vitamin C), while vitamin A is also light-sensitive.

As discussed in Chapter 5, the shelf-life of pasteurized milk is very much determined by the heat treatment applied, in terms of achieving the target temperature (72–74°C) and holding it there for the required time (15–30 seconds), which results in inactivation of pathogenic bacteria and a reduction in the numbers of spoilage bacteria, so as to get a reasonable refrigerated shelf-life. However, this presupposes that the levels of bacteria are determined and controlled entirely by this heating step, and that there is no reintroduction or contamination by spoilage bacteria after the heat treatment. Indeed, for many decades one of the greatest influences on the shelf-life and quality of pasteurized milk was what is called post-pasteurization contamination (PPC), which may occur during the packaging process itself. Glass bottles were especially susceptible to such recontamination problems.

Because of the disadvantages of using glass bottles for packaging of milk, in many countries the most common package for liquid milk today (whether pasteurized or UHT-treated) is the Tetra Pak, which was developed by the Swedish company of the same name in the 1950s and first used for the packaging of UHT-treated milk in Switzerland in 1961. The key to ensuring a long shelf-life is ensuring that, once heat treatment has rendered the milk microbiologically sterile, no recontamination (PPC) can occur. The Tetra Pak was originally introduced for just this function, as it was designed for aseptic packaging, which means packaging a sterile product in a sterile package in a sterile environment.

Today it has replaced glass bottles for the packaging of pasteurized milk in most countries.

The Tetra Pak is made primarily from paperboard, which constitutes 75% of the carton; the remainder is polyethylene (20%) and aluminum (5%). These three materials are layered using heat and pressure to form a six-layered structure called a laminate that protects the contents from light, oxygen, air, and moisture. The paperboard provides strength and printability, the polyethylene protects against moisture (and enables the paperboard to stick to the foil), and the aluminum foil protects against oxygen and light; together the materials deliver the protection required for the product. Both the outer and inner surfaces of the Tetra Pak are polyethylene.

The material for a Tetra Pak is typically supplied on a roll which, in the packaging machine, is formed into the familiar shape. The bottom is then sealed and the product added, with the lid being sealed before the package exits the sterile filling chamber; this is known as a form-fill-seal system. Another option involves filling followed by sealing of prefabricated cartons.

The key step in ensuring the long shelf-life of UHT-treated milk is aseptic packaging. One of the key advantages of the Tetra Pak filling system is its ability to be sterilized by a number of means, including irradiation, pulsed light, UV radiation, plasma, steam, hot air, or chemical sterilants such as hydrogen peroxide (which may be applied as a spray, dip, or rinse) or peracetic acid.

As UHT milk is frequently used in mini-pots or sachets for addition to tea or coffee (especially in planes, trains, and hotels), it is also packaged in small cups, sachets, or pouches. The cups are made of a tough form of polystyrene called high impact polystyrene (HIPS), which is formed into cups, filled aseptically, and sealed.

In some countries, milk is packaged in plastic bottles or jugs, which may be made of blow-molded high-density polyethylene (HDPE), a hard opaque plastic that is lightweight but strong, and will typically have a low-density polyethylene (LDPE) cap and printed label. Clear plastic bottles made from materials such as PET may also be used, but transparent packages have some of the same disadvantages as glass in terms of light transmission. This has been countered, in some cases, by incorporating a white pigment, called titanium dioxide, into the plastic; however, this is now discouraged due to concerns over the health implications of titanium. Addition of a green pigment to the plastic to block the specific wavelengths of light that damage the product is an effective alternative.

## Milk powder

In the case of milk powder, a key goal is to prevent moisture being absorbed during storage. The addition of moisture to the powder can result in the uncontrolled crystallization of lactose, which leads to the undesirable phenomenon called caking: the formation of large interlocked crystals. When that happens, the once free-flowing powder transforms into a large, solidified mass, which is very difficult to sample and handle.

Thus, the packaging selected for milk powder should be a good moisture barrier, and so a cardboard bag or sack (for strength, printability, and exclusion of light, to prevent oxidation) lined with a plastic film like PE is common. Such bags typically, when filled, contain 25 kg of powder. The bag might also be flushed with chemically inert nitrogen to deplete any oxygen-containing air present.

The presence of gas can actually be critical in milk powder, as powders contain significant amounts of gas (air or nitrogen) and spaces into which water can easily permeate on later reconstitution, making that process significantly easier. So, while this increases the volume of the powder (such powder is referred to as having a low bulk density), it favors more convenient later use.

Cans may also be used for packaging dry dairy products, such as infant formula. These are typically gas-flushed (with the oxygen-rich air being replaced by an inert gas such as nitrogen) to ensure no oxidation of sensitive nutrients occurs during long-term storage. Sacks of milk powder may also be gas-flushed before sealing, to prevent oxidation.

## Butter and spreads

The packaging considerations for butter and spreads, although they are two products often found together on the market shelf, are somewhat different. Both need a light-proof package (to prevent fat oxidation). Hence foil-lined or translucent plastic materials are common, as these are resistant to the solvent effects of many fats that might soften many other materials (e.g., paper must be made greaseproof to prevent this from happening). Fat can also readily absorb many compounds from the atmosphere, which could lead to off-flavor production, and so the package must protect against this.

However, butter, at least at the low temperatures at which is it typically transported and stored, is normally sufficiently solid to maintain its own shape, and does not need a strong package, so plastic-lined aluminum foil is typical. Spreads, though, being much softer, would deform easily in such a package and need a rigid tub to physically protect and support them. Packages should also be odor impermeable, as milk fat is a good solvent that can readily absorb off-flavors and off-odors from other foods stored alongside the butter.

In industrial practice, immediately after manufacture butter is often packaged in 25-kg blocks in plastic-lined cardboard boxes and cooled and solidified in those. At this point, the freshly made butter has a temperature of around 12°C.

Subsequent cooling to refrigeration temperature (4°C) results in significant hardening of the product, as additional fat crystallizes (continuing the process described in Chapter 8). Some weeks later, the butter will be mechanically worked and kneaded to physically soften it and then transferred to smaller (250–500 g) consumer packages for sale. So, in this case, the packaging operation represents a key step in modifying and optimizing the characteristics of the product.

Specialty or artisan butter may be packaged and wrapped by hand in greaseproof paper, which, as the name suggests, is waxed to protect it from the solvent properties of milk fat. Wrappers should also prevent loss or gain of moisture; moisture loss from the surface of butter can result in a visible defect called mottling.

## Yoghurt and ice cream

The key property of packages for both yoghurt and ice cream is shape, so that the contents can readily be scooped out by consumers; they come in a range of sizes and configurations. Yoghurt containers may even have different compartments, one for the yoghurt itself and the other for a flavoring component such as chocolate, biscuit, or fruit pieces to be mixed through the yoghurt before consuming it.

Typically, thermo-formed HIPS cups or HDPE tubs with aluminum/plastic laminate lids (for yoghurt) or rigid plastic lids (for ice cream) are used. Glass may also be used, especially for premium products.

In the case of probiotic products, a good oxygen barrier is necessary, as many of these bacteria are susceptible to oxygen and prefer an oxygen-free (anaerobic) environment. Ensuring that sufficient numbers of cells survive to the point of consumption (to allow the products to be declared as probiotic) is highly dependent on the packaging used, so multilayer packages are often used. Sometimes, pouches of materials that scavenge oxygen can also be included in the package lid for this purpose.

## Evaporated and sweetened condensed milk

One of the oldest forms of food package is the aluminum can, which was developed originally for the purpose of containing food during high temperature sterilization processes, and even today canned foods are associated with very long shelf-life products. In most countries, the principal dairy products still found in cans are evaporated and sweetened condensed milk. These products are typically used in bakery applications, and so the significant color and flavor changes induced by the high temperatures applied in the sterilization of the product are less undesirable than they would be for most other dairy products.

## Cheese

Cheese is a generic term covering a wide range of diverse products, with very different textures, flavors, structures, and ripening requirements. Some hard cheeses have a strong solid structure, and thus minimal requirements for packaging, whereas the packaging requirements for a very soft cheese—to ensure it is not squashed during storage and transport—are very demanding. In addition, most types of cheese undergo significant periods of ripening after production, during which biochemical reactions take place that determine the ultimate characteristics of the cheese. The choice of package during this stage, and in particular, the passage of moisture and gases into (and sometimes out of) the package are critical factors.

Through much of its history (i.e., in the pre-plastic era), cheese was packaged in various cloth or paper-based wrappings or, often, left unwrapped to develop a dry and impermeable surface layer (rind), which acted as a form of natural packaging.

Today, in the case of one of the cheeses made on the largest industrial scale, Cheddar, the manufacturing process typically ends in the cheese being formed into 25-kg blocks in a machine called a block-former, after which it is immediately vacuum-packed in polyamide or PE film, for later subdivision into consumer packages.

In the case of traditional Dutch-type cheeses, such as Edam and Gouda, the cheese is coated after brining and drying in a colored plastic coat or hard paraffin wax, perhaps in different colors for different varieties or manufacturers. This coating is typically applied in a number of coats or layers, and might include antifungal agents, like natamycin. Different varieties can then be identified by the color of the coating, red for Edam and yellow for Gouda.

In the case of cheeses that need to "breathe" during ripening, where the growth of surface microorganisms in smear-ripened cheese (e.g., Tilsit) or mold (e.g., Brie and Camembert) is critical to the character of the cheese, the permeability of the package to oxygen and carbon dioxide are critical considerations. In such cases, deliberately porous materials like foil laminates or perforated plastics may be used.

Protection for soft cheeses may be provided by placing the cheese in a soft package or wrapping into an outer protective box made of card, light strips of timber, or perhaps thermoformed PS or PVC. Cottage cheese might be packaged in a carbon dioxide-enriched atmosphere (carbon dioxide being an antimicrobial agent) in PS tubs.

In the case of cheese where an open structure is critical, such as Emmental, the cheese will actually increase in volume during ripening, so it is not usually packaged until after the large eyes have formed. For such cheese types, modified atmosphere packaging may be used: a tailored gas mixture is introduced into the package to create what is essentially a flexible protective shape, like a pillow.

One product related to cheese, processed cheese, is defined by its flexibility in production: a molten mix is poured into, and solidifies in, a range of different sized packages and shapes, taking their size and shape. This feature of its manufacture can be exploited to generate product formats including sausage, triangle, block, or slice shapes. So, typical packaging materials might be LDPE sleeves for slices, laminated foil for blocks or triangles, or thermoformed tubes for very soft spreadable products. To prevent soft processed cheese slices from sticking to each other or to the package, modified atmosphere packaging is used, with the package typically containing a mixture of carbon dioxide and nitrogen.

## Active packaging

In recent years, there has been a significant evolution in the understanding of how packaging can interact with and enhance the quality of the food with which it is in contact. Terms like "intelligent" or "active" packaging have become common and refer to packaging that possibly provides information about the state of the food contained therein or actively interacts with the food in a way that helps maintain quality and safety.

In terms of the type of information that a package might monitor and reflect, one of the key factors that influences the shelf-life of most dairy products is temperature. Even a few degrees fluctuation above refrigeration temperature could have profound implications for the shelf-life of pasteurized milk, as the growth rates of bacteria are highly influenced by temperature. However when a consumer obtains milk, they have no idea of its temperature history and whether it might have been stored previously at a higher temperature—the use-by or other indicative dates displayed relate to storage under ideal conditions.

So, if a package bore a sensor or meter, perhaps a colored strip, that showed if the milk had warmed significantly at any point, it would provide additional information on the quality and usability of the milk. A simple way to do this might be to have a tiny sachet attached to the outside of the package with bacteria, sugar, and a pH indicator. In the cold, the bacteria do not grow, but if a milk container was left out of the fridge, they would, and thereby break down the sugar and produce acid, leading to a change in the indicator dye (perhaps from green to red). The same effect could be achieved by using temperature-sensitive inks or digital temperature loggers. Such intelligent packaging gives consumers information about the history of the product and is likely to become more common in the future, as it provides far more dynamic and accurate information than static shelf-life dates.

Packages can also contain microchips with radio-frequency identifiers (RFIDs), which allow much more information about the product and its history to be stored and accessed. This allows packaging to be much more functional and informative to users, but obviously adds cost to a food product.

Active packaging is already used in other food categories, e.g., a sachet in a meat package that absorbs oxygen during storage (called oxygen scavengers), or packaging films impregnated with a substance like a silver salt that kills microorganisms growing in the food. Materials that can absorb moisture, carbon dioxide, odors, and other substances, or can inhibit oxidation through incorporation of antioxidants such as vitamin E, have also been developed for use in food packaging.

Finally, one area of interest in recent years has been the development of edible packaging, so the package becomes a consumable part of the food product. While applications for dairy products are not well established (except for the use of polysaccharide-based coatings for cheese), interestingly, dairy materials (such as milk proteins) are used as elements of edible packages for other products.

# 15
# Challenges in the Modern Dairy Sector

At the time of writing, the place of dairy products within the global food system is facing challenges unprecedented in its history. Although dairy products have been made and consumed around the world since ancient times and represent a key element of daily nutrition for billions of people, societal views on dairy products are under question on many fronts, particularly with the rise of veganism. In this short concluding chapter, some of the challenges being faced by the dairy sector today will be considered and some possible future directions examined.

## The resistance to dairy products

In the first decades of the 21st century, veganism—diets that exclude all animal-related products, including meat and dairy products—was a rising trend in many countries. In 2021, it was estimated that 1% of the world's population was vegan (although in some regions such as Africa and Asia this had long been the case for cultural reasons).

One key driver for the rise in veganism relates to ethical considerations around animal welfare, especially in the light of increasingly large-scale farming with enormous herds in some countries. It should be noted, however, that the average herd size in the world is actually 2–3 cows, due to the huge number of small farmers in the developing world.

Increasing concern around the animal welfare implications of milk production has placed greater focus on this aspect. Milk processors and farmers today strive to ensure the highest standards for their cows, and such aspects as cows being allowed extensive outdoor grazing on grass (as in Ireland and New Zealand) are increasingly valued. In addition, companies increasingly highlight their ethical treatment of animals in their communications and marketing.

A second factor causing people to opt for a vegan diet is the environmental impact of dairy (and meat) production. A related major consideration among societies at this time is the climate change emergency and global warming. Agriculture, including dairy production, is seen as making a significant contribution to the production of greenhouse gases such as methane, and a key driver

*From Farm to Table*. Alan Kelly, Patrick Fox, and Tim Cogan, Oxford University Press.
 DOI: 10.1093/9780197581025.003.0015

in the global dairy sector is the need for more sustainable farming practices, which cause less environmental damage and do not accelerate global warming. These include reduced use of pesticides and fertilizers, adoption of new types of feed (especially use of grass for milk production), reducing losses to the environment, and improving the efficiency of farming systems.

## The emergence of dairy alternatives

The avoidance (or reduction) of the consumption of dairy products due to these considerations has led to significant research and product development to allow consumers to have dairy-like contributions in their diet, without actually consuming traditional dairy products. Milk is unquestionably one of the most nutrient-rich foods, and so avoiding or replacing milk in the diet is a complex proposition, which, in the case of veganism, can result in a requirement to take specific supplements to make up the shortfall in intake of critical nutrients.

The search for milk alternatives continues to accelerate, nonetheless, and one area of scientific effort has seen a number of companies (particularly small biotechnology firms) exploring the production of "milk" without the involvement of cows or other mammals. In one approach, the genes that encode production of milk proteins have been engineered into microorganisms like yeast to allow milk proteins to be produced in vats in so-called "precision fermentation" processes—as has been used extensively in recent decades to produce chymosin for cheese-making, avoiding use of the original historical source, i.e., calf stomachs. These proteins could then be combined with sugars, lipids, minerals, and vitamins to produce a milk-like material.

However, as outlined in Chapter 2, milk proteins and the structures they form that give dairy products their specific properties (like the casein micelle) are exceptionally complex and replicating the assembly of a family of proteins into these structures is a daunting scientific challenge. This is especially so, as very minor aspects of milk protein structure (such as the attachment of phosphate groups and sugars to the protein molecules) are very challenging to replicate in systems like yeast but are critical to yielding a product with genuinely milk-like properties. The other milk proteins, i.e., whey proteins, are simpler to produce, and some products such as ice cream containing laboratory-produced milk proteins are made in the United States.

Other companies are reportedly working on growing mammary cells in culture flasks surrounded with suitable media rich in raw materials that will allow them to secrete milk in vitro, rather than in vivo in the mammary gland. There is particular interest in producing human milk in this way, to create an alternative to infant formula for mothers who cannot, or choose not to, breastfeed.

Other approaches to producing alternatives to dairy products include the use of other food ingredients to produce products with a texture, appearance, and flavor similar to that of dairy products. For example, in many countries vegan cheese in block, spread, or sliced form is available. Such products are typically produced by a process like that used for processed cheese, but frequently rely for their structure on agents like starch rather than the protein structures found in cheese and yoghurt. As a result, they are generally less nutritious than their dairy equivalents in terms of providing key nutrients such as protein.

In many countries, a range of milk alternatives is available, including oat, almond, soya, and even potato milks. Technically, these should not be referred to as milks, as that term relates to the mammary secretion of animals. In some countries the terminology used for these products is becoming more closely scrutinized, but they have become a recognized alternative to bovine milk and can frequently be found in coffee shops as a (typically more expensive) alternative base or additive for coffee drinks.

To recognize the desire of some consumers not to exclude dairy products entirely from their diet, but reduce their intake, perhaps for reasons related to concerns around sustainability, a range of so-called hybrid products has emerged. These include blends of milk or milk-derived ingredients with plant-derived proteins—from peas, lentils, or cereals. A number of successful products of this type have been launched, but the scientific challenges of ensuring desirable flavor profiles, textures, and shelf stability of such products are not trivial.

## New approaches to processing

As outlined in previous chapters, dairy products represent some of the oldest examples of the application of food processing principles that have since become commonplace for other food products, such as salting, dehydration, fermentation, and heat treatment.

As well as the societal trends mentioned above that have focused on dairy products, there is consumer resistance to methods of processing that result in significant alterations to the raw material. Freshness and naturalness are attributes that are greatly valued by consumers, and there is suspicion of foods that are seen as "processed" or "ultra-processed," the latter term referring to food products that have large numbers of ingredients, especially those isolated using complex processes (such as those used for milk proteins as described in Chapter 10).

There is thus a trade-off between preservation and avoidance of undesirable changes and the extent to which microorganisms and enzymes are inactivated. The difference in consumers' perceptions of the quality and desirability of pasteurized and UHT milk is a good example of this.

This has created a goal of "minimal processing": the final product appears as little changed as possible from the fresh unprocessed equivalent, especially in sensory and nutritional terms. New processing technologies have emerged, as a result, that can achieve such aims while, critically, ensuring that the product is safe for consumption (the key role of processing and food additives in ensuring food safety is not always a prominent part of debates on simpler approaches to food processing). Minimal processing requires technological solutions that can achieve the positive outcomes of heat treatment without the negative consequences and provide the best of both worlds—fresh yet safe and stable.

To take one example of this, while Louis Pasteur was developing the science of the process that today bears his name, on the other side of the Atlantic another researcher was experimenting with a quite different approach to making food last longer. Bert Hite, of the West Virginia Agricultural and Forestry Experiment Station, was experimenting in the 1890s with a machine within which he could subject materials to pressures such as those that might be encountered deep undersea or underground. He showed that food materials like milk could indeed be made to last longer following such treatment, although no heat was involved, and even did some tests that showed that pathogenic bacteria could be killed by pressure.

In modern high-pressure (HP) treatment processes, food products are subjected to enormous pressures, several times higher than those found at the deepest parts of the ocean, like the Mariana trench. This is achieved by submerging the food, in a suitable package, in a fluid within a strong metal chamber. The fluid is then compressed using pumps or a piston to generate the pressures required. The product is held, usually at ambient temperature, at the target pressure for several minutes, before the pressure is released and the product removed.

HP treatment is generally a batch process, or a stop/start process very much disliked by the food industry, due to the amount of down (non-processing) time involved and the labor required in loading and unloading. This, along with the relatively small scale of HP treatment systems (large machines still have a chamber size of only a few hundred liters in volume) and their very high cost (systems can cost several million dollars), remain barriers to the uptake of HP processing by the food industry.

The high cost means that HP treatment makes sense only when the value added by the process justifies the initial high capital outlay for equipment. In some countries, this challenge has been overcome by what are called toll-processing operations: a company buys a high-pressure processing system and then rents out time for its use to other processors, who cannot purchase such expensive technology themselves.

HP processing is currently being used commercially in many countries for products such as fruit juices, meat, avocado, and shellfish, due to its ability to kill bacteria (many of which are very sensitive to the pressures applied) with little impact on flavor or nutritional quality.

In terms of dairy products, HP treatment has been proposed for such applications as accelerating or decelerating the ripening of cheese (the costly and dynamic process whereby the characteristic flavor and texture develop over a period of months) or rendering cheese microbiologically sterile before its addition to other food products, like sandwiches.

There is another example of the use of pressure for dairy applications for a very specific application, for which the low volume and high cost of HP was less of a concern due to the specialized and valuable nature of the product in question, i.e., colostrum. Colostrum is highly enriched in biologically significant components, especially proteins (immunoglobulins), which boost the immune system and promote health and development.

In many European countries, colostrum is not regarded as a common food for human consumption, but in some Asian countries it is highly valued for its health-promoting properties. But processing it for such use has always been challenging, as, like any milk-based material, it must be treated to make it safe and extend its shelf-life, usually by heat treatment. However, heat treatment also denatures and changes the structure and function of proteins, which impairs the very health-promoting properties of colostrum. It was shown, in New Zealand, that high-pressure treatment of colostrum, particularly if combined with mild acidification, could result in elimination of microbes but retention of the desired biological functions.

There is a growing interest in the consumption of raw milk, due to concerns around a perceived (although scientifically questionable) loss of flavor and nutritional benefit due to pasteurization. In Australia, around 2018, a company called Made by Cow launched HP-treated raw milk, though it is actually described, using a term which has become quite common elsewhere for products like juice, as "cold-pressed." It cannot be described as pasteurized because it has not received a heat treatment equivalent to pasteurization, but it has purportedly attained a state of microbiological quality significantly greater than that of raw milk. This is a strong and potentially controversial claim to make, but the relevant Australian food safety bodies deemed the product safe for sale, and allowed the company to describe the product as "100% safe to drink." The milk was sold for around twice the cost of regular milk, and is very creamy.

One key to the microbiological quality of the product appears to be careful selection of the raw milk, which is poured into plastic bottles and then subjected to high pressures. This milk comes from cows and farms producing hygienic raw milk of the highest possible quality, and thus the milk starts off with the lowest possible microbial load. This is a nice illustration of an important principle of any type of food processing, which is that the best way to produce a high-quality product is to start off with a high-quality raw material.

There is interest today in a number of countries in launching such products, which represent a safe alternative to raw milk for those who prefer to avoid pasteurized milk. But the ultimate acceptability of this concept will depend on the strictures and views of local food safety authorities, due to the potential danger of raw milk that has not been handled properly.

So, overall, after a slow commercial start, during which a significant research effort focused on the area, HP-treated dairy products are beginning to appear on shelves around the world. Due to the challenges regarding cost and the scale of the equipment required to apply the process, this is likely to remain an option mostly for small-volume, high-value food products, rather than becoming a widespread bulk commodity processing technology.

HP treatment is not the only technological alternative to heat that has been explored to produce high-quality, fresh, minimally processed food products. Another principle which has been around for a surprisingly long time is the use of electricity, specifically very short pulses of electricity passed through a material, such as food, which is the basis of a process called pulsed electric field (PEF) treatment.

It is known that such rapidly alternating electric fields interfere with key parts of the cell walls of bacterial cells, part of the function of which is to control the movement of materials into and out of the call. Electric fields can overcome such systems and result in holes being punched in the protective outer layers of the cells, which essentially allow critical cell contents to leak out, leading ultimately to the death of the cell.

This kind of effect has led to many of the current commercial applications of PEF for food, where the ability to permeabilize cells can be used to enhance extraction of valuable components from cells, such as sugars, flavors, or nutrients. In some countries, a major commercial application of PEF is to enhance the peeling of potatoes for applications like French fries.

PEF treatment for milk was demonstrated over 100 years ago, when a process called ElectroPure was first reported—a paper on the technology appeared in the *Journal of Dairy Science* for September 1919—but, as for HP treatment, a lack of suitable equipment meant that this was little exploited for most of the 20th century. Again, as for pressure, the first commercial applications that have appeared in recent decades were for non-dairy products, such as fruit juice, for which PEF offered some of the same advantages as HP treatment, such as enhanced safety and shelf-life, combined with good retention of sensory and nutritional quality.

In recent years, commercial systems for continuous-flow treatment of milk using PEF have been reported, and are available, but the uptake is low and it will be interesting to see how this technology will find applications in the dairy sector in the future.

Other new technologies that are being explored for the processing and preservation of milk include scaled-up versions of microwave processing, very familiar from domestic food preparation but perhaps surprisingly little used to date on an industrial scale, and homogenization at much higher pressures than traditionally used (such that the forces encountered kill the micro-organisms), and it will be interesting to see in the coming years how many of these technologies are used to produce new generation of high quality milk products.

Overall, while dairy researchers today continue to explore both the underlying science of milk and dairy products, which retain surprises and new developments even after decades of study, dairy processors seek to apply new approaches to ensure the production of a wide diversity of high-quality, safe, stable, and nutritious products from that one single raw material. The market for consumers of dairy products, however, and the demands and expectations of consumers, as well as external economic and environmental factors, will continue to have a great influence on this sector, and will no doubt shape the dairy sector of the future.

# Notes


## Chapter 2

1. Evershed, R.P.; et al. Dairying, disease and the evolution of lactase persistence in Europe. *Nature*, **2022**, 608, 336–345. https://doi.org/10.1038/s41586-022-05010-7.
2. Early views and research on casein micelles was reviewed by Fox, P. F.; Brodkorb, A. The casein micelle: historical aspects, current concepts and significance. *Intern Dairy J.*, **2008**, 677–684. http://dx.doi.org/10.1016/j.idairyj.2008.03.002.
3. For general information on the vitamins and, for specific aspects in relation to milk and dairy products, including stability during processing and storage, the reader is referred to a set of articles in Roginski, H.; Fuquay, J. W.; Fox, P. F. Encyclopedia of Dairy Sciences. 2003. Academic Press, Amsterdam. Fuquay, J. W.; McSweeney, P. L. H.; Fox, P. F. Encyclopaedia of Dairy Sciences. 2nd Edn. **2011**. Elsevier.

## Chapter 4

1. Schleifer, K.H.; Kilpper-Balz, R. Molecular and chemotaxonomic approaches to the classification of streptococci, enterococci and lactococci: a review. *Syst. Appl. Microbiol.*, **1987**, *10*, 1–9.
2. Schleifer, K.H.; Kraus, J.; Dvorak, C.; Kilpper-Balz, R.; Collins, M.D.; Fischer, W. Transfer of *Streptococcus lactis* and related streptococci to the genus *Lactococcus* gen. nov. *Syst. Appl. Microbiol.*, **1985**, *6*,183–195.
3. Zheng, J.; Wittouck, S.; Salvetti, E.; et al. A taxonomic note on the genus *Lactobacillus*: Description of 23 novel genera, emended description of the genus *Lactobacillus* Beijerinck 1901, and union of Lactobacillaceae and Leuconostocaceae. *Int. J. Syst. Evol. Microbiol.*, **2020**, *70*, 782–2858.
4. Franz, C.M.A.P.; Stiles, M.E.; Schleifer, K.H.; Holzapfel, W.H. Enterococci in foods—a conundrum for food safety. *Int. J. Food Microbiol.*, **2003**, *88*, 103–122.
5. Fortina, M.G.; Ricci, G.; Borgo, F; et al; A survey on biotechnological potential and safety of the novel *Enterococcus* species of dairy origin, *E. italicus*. *Int. J. Food Microbiol.*, **2008**, *123*, 204–211.
6. Kelly, W.J.; Ward, L.J.H.; Leahy, S.C. Chromosomal diversity in *Lactococcus lactis* and the origin of dairy starter cultures. *Genome Biol. and Evol.*, **2010**, *2*, 729–744.
7. Jenkins, J.K.; Harper, W.J.; Courtney, P.D. Genetic diversity in Swiss cheese starter cultures assessed by pulsed field gel electrophoresis and arbitrarily primed PCR. *Letters in Applied Microbiology*, **2002**, *36*, 423–427.

## Chapter 5

1. Robertson, R.; Cerf, O.; Condron, R.; Donaghy, J. Review of the controversy over whether or not *Mycobacterium avium* subsp. *paratuberculosis* poses a food safety risk with pasteurised dairy products. *IDJ*, **2017**, *73*, 10–18.

## Chapter 6

1. Yang Y.; Shevchenko A.; Knaust A.; Abuduresule I.; Li W.; Hi X.; Wang C,; Schevenko A. Proteomics evidence for kefir dairy in early bronze age China. *J. Arch. Sc.*, **2014**, *45*, 178–186.
2. Cotter, P.D.; Hill, C.; Ross, R.P. Bacteriocins: developing innate immunity for food. *Nature Reviews, Microbiology*, **2005**, *3*, 777–788. Drider, D.; Fimland, G.; Hechard, Y.; et al. The continuing story of Class IIa bacteriocins. *Microbiol. Mol. Reviews*, **2006**, *70*, 564–582. Quinto, E.J.; Jimeniz, P.; Caro, I.; et al. Probiotic lactic acid bacteria: a review. *Food and Nutrition Sciences*, **2014**, *5*, 1765–1775.

3. Further information on manufacture of the different varieties can be found in the relevant chapters of Fox, P. F.; McSweeney, P. L. H.; Cogan, T. M.; Guinee, T. P. *Cheese: Chemistry, Physics and Microbiology*. Volume 2, 3rd edn. **2017.**
4. Holle, M. J.; Ibarra-Sanchez, L. A.; Stasiewicz, M. J.; Miller, M. J. Microbial analysis of commercially available US Queso Fresco. *J. Dairy Sci.*, **2018**, *101*, 7736–7745.
5. Khan, S.U.; Pal, M.A. Paneer production: a review. *J. Food Sci. Technol.*, **2011**, *48*, 645–660; Kumar, S.; Rai, D.C.; Niranjan, K.; Bhat, Z.F. Paneer—An Indian soft cheese variant: a review. *J. Food Sci. Technol.*, **2014**, *51*, 821–831.

## Chapter 7

1. International Dairy Federation Standard 163:1992.
2. Prado, M.R.; Blandon, L.M.; Vandenberghe, L.P.S; et al. Milk Kefir: composition, microbial cultures, biological activities and related activities. *Front. Microbiology,* **2015,** *6*, Article 1177; Nejati, F.; Junne, S.; Neubauer, P. A big world in a small grain: a review of natural milk kefir starters. *Microorganisms,* **2020,** *8*, 192–202.
3. Walsh, A.M.; Crispie, F.; Kilcawley, K.; et al. Microbial succession and flavor production in the fermented dairy beverage Kefir. *M. Systems,* **2016,** *1*, e00052–16.
4. Rea, M.C.; Lennartsson, T.; Dillon, P.; Drinan, F.D.; Reville, W.J.; Heapes, M.; Cogan, T.M. Irish kefir-like grains: their structure, microbial composition and fermentation kinetics. *J. Appl. Microbiol.,* **1996,** *81*, 83–94.
5. Guo, L.; Ya, M.; Guo, Y.G. et al. Study of bacterial and fungal community structures in traditional koumiss from Inner Mongolia. *J. Dairy Sci.*, **2019**, *102*, 1972–1984.
6. Kahala, M.; Maki, M.; Lehtovaara, J. M.; Katiska, M.; Juuruskorpi, M.; Joutsjoki, V. Characterization of starter lactic acid bacteria from the Finnish fermented milk product villi. *J. Appl Microbiol.*, **2008,** *105*, 1929–1938.
7. Roginski, H. Fermented milks: Nordic fermented milks, in Fuquay, J.; Fox, P.F.; McSweeney, P. 2002. *Encyclopaedia of Dairy Sciences*, Vol 2. 2nd edn. Academic Press, 496–502.
8. Abd El-Salem, M.H. 2002. Middle Eastern fermented milks, in Fuquay, J.; Fox, P.F.; McSweeney, P. 2002. *Encyclopaedia of Dairy Sciences*, Vol 2. 2nd edn. Academic Press, 503–506.
9. Akuzawa, R.; Miura, T.; Surono, I. S. Asian fermented milks, in Fuquay, J.; Fox, P.F.; McSweeney, P. 2002. *Encyclopaedia of Dairy Sciences*, Vol 2. 2nd edn. Academic Press, 507–511.
10. Vasiljevic, T.; Shah, N.P. Probiotics–from Metchnikoff to bioactives. *International Dairy Journal,* **2008,** *18*, 714–728.
11. Bourrie, B.C.T.; Willing, B.P.; Cotter, P.D. The microbiota and health promoting characteristics of the fermented beverage Kefir. *Frontiers in Microbiol.,* **2016**, *7*, 1–17.

## Chapter 8

1. Cronin, T.; Downey, L.; Synott, C.; McSweeney, P.L.H.; Kelly, E.P.; Cahill, M.; Ross, R.P.; Stanton, C. Composition of ancient bog butter. *Int. Dairy J.,* **2007**, *17*, 1011–1020.
2. Conway, V.; Gauthier, S.F.; Pouliot, Y. Buttermilk: much more than a source of milk phospholipids. *Anim. Front.,* **2014**, 44–71.

## Chapter 11

1. https://www.smithsonianmag.com/smart-news/george-washington-liked-ice-cream-so-much-he-bought-ice-cream-making-equipment-capital-180950316/#:~:text=Cool%20Finds-,George%20Washington%20Liked%20Ice%20Cream%20So%20Much%20He%20Bought%20Ice,Making%20Equipment%20for%20the%20Capital&text=In%20the%2018th%20century%2C%20ice,Smith%20Jr.

# Further Reading


## Dairy Science (General)

McSweeney, P.L.H., and McNamara, J.P. (eds.) (2021) *Encyclopedia of Dairy Sciences.* Academic Press Inc., Cambridge, Massachusetts.

## Dairy Chemistry

Fox, P.F., Uniacke-Lowe, T., McSweeney, P.L.H., and O'Mahony, J.A. (2015) *Dairy Chemistry and Biochemistry.* Springer, New York.

McSweeney, P.L.H., Fox, P.F., and O'Mahony, S.A. (2020) *Advanced Dairy Chemistry Vol 2: Lipids.* Springer, New York.

McSweeney, P.L.H., O'Mahony, S.A., and Fox, P.F. (2013) *Advanced Dairy Chemistry Vol 1A and B: Proteins.* Academic Press, Cambridge, Massachusetts.

McSweeney, P.L.H., O'Mahony, S.A., and Kelly, A.L. (2022) *Advanced Dairy Chemistry Vol 3: Lactose, Proteins and Minor Constituents.* Springer, New York.

Singh, H., Boland, B., and Thompson, A. (2014) *Milk Proteins: From Expression to Food.* 2nd edn. Academic Press, Cambridge, Massachusetts.

## Dairy Processing (General)

Bylund, G. (2015) *Dairy Processing Handbook.* Tetra Pak Processing Systems AB, Sweden.

Datta, N., and Tomasula, P.M. (eds.) (2015) *Emerging Dairy Processing Technologies: Opportunities for the Dairy Industry.* Wiley-Blackwell, Chichester, UK.

Early, R. (2007) *Technology of Dairy Products.* Springer, New York.

Varnam, A., and Sutherland, J.P. (1994) *Milk and Milk Products: Technology, Chemistry and Microbiology.* Springer, New York.

Walstra, P., Wouters, J.T.M., and Geurts, T.J. (2006) *Dairy Science and Technology.* 2nd edn. CRC/Taylor and Francis, Boca Raton, Florida.

## Dairy Microbiology

Doyle, M.P., and Beuchat, L.R. (2007) *Food Microbiology: Fundamentals and Frontiers.* 3rd edn. ASM Press, Washington DC.

Marth, E.H., and Steele, J. (2001) *Applied Dairy Microbiology.* 2nd edn. Marcel Dekker, New York.

Robinson, R.K. (ed.) (2002) *Dairy Microbiology Handbook.* 3rd edn. John Wiley & Sons, New York.

## Cheese

Eck A., and Gilles J.C. (2000) *Cheese-making: From Science to Quality Assurance.* Technique et Documentation (Lavoisier), Paris.

Fox P.F., Guinee T.P., Cogan T.M., and McSweeney P.L.H. (2017) *Fundamentals of Cheese Science.* 2nd edn. Springer, New York.

Kosikowski F.V., and Mistry V.V. (1997) *Cheese and Fermented Milk Foods,* Vols. 1 and 2. 3rd edn. FV Kosikowski LLC, Westport, CT.

Law, B.A. (1997) *Microbiology and Biochemistry of Cheese and Fermented Milks.* 2nd edn. Chapman & Hall, London.

Law, B.A. (1999) *The Technology of Cheese-making.* CRC Press, Boca Raton, Florida.

McSweeney, P.L.H. (2007) *Cheese Problems Solved.* Woodhead Publishing, Cambridge, UK.

McSweeney, P.L.H., Fox P.F., Cogan, Timothy M., and Guinee, Timothy P. (2017) *Cheese: Chemistry, Physics and Microbiology*. Vols. 1 and 2. 3rd edn. Elsevier Academic Press, London.

Robinson, R.K., and Wilbey, R.A. (1998) *Cheese-making Practice*. 3rd edn. Aspen Publishers, Gaithersburg, MD.

Tunick, M.H. 2014. *The Science of Cheese*. Oxford University Press, New York.

## Chocolate

Beckett, S.T. (2000) *The Science of Chocolate*. Royal Society of Chemistry, Cambridge.

Beckett, S.T. (ed.) (1988) *Industrial Chocolate Manufacture and Uses*. Blackie & Son, Glasgow.

Beckett, S.T., Fowler, M.S., and Ziegler, G.R. (eds.) (2017) *Industrial Chocolate Manufacture and Uses*. 5th edn. John Wiley & Sons, New York.

Miniifie, B.W. (1989) *Chocolate, Cocoa and Confectionary*. 2nd edn. Avi Publishing Co. Inc., Westport, CT.

## Ice cream

Goff, H.D., and Hartel, R.W. (2013) *Ice Cream*. Springer, New York.

Marshall, R.T. (2003) *Ice Cream*. Springer, New York.

## Yoghurt

Tamine, A.T., and Robinson, R.K. (2007) *Yoghurt: Science and Technology*. Woodhead Publishing, Cambridge, UK.

## Milk Powders and Ingredients

Corredig, M. (ed.) (2009) *Dairy-derived Ingredients: Food and Neutraceutical Uses*. Woodhead Publishing, Oxford, UK.

Piseck, J. (2012) *Handbook of Milk Powder Manufacture*. GEA Process Engineering SA, Copenhagen, Denmark.

Tamime, A.T. (2009) *Milk Powders and Concentrated Dairy Products*. Wiley-Blackwell, Hoboken, New Jersey.

## Infant Formula

Guo, M. (2014) *Human Milk Biochemistry and Infant Formula Manufacturing Technology*. Woodhead Publishing, Cambridge, UK.

# Index

*For the benefit of digital users, indexed terms that span two pages (e.g., 52–53) may, on occasion, appear on only one of those pages.*

Figures are indicated by an italic *f* following the page number.